ISBN 978-1-331-97661-5
PIBN 10263291

This book is a reproduction of an important historical work. Forgotten Books uses
state-of-the-art technology to digitally reconstruct the work, preserving the original format
whilst repairing imperfections present in the aged copy. In rare cases, an imperfection in
the original, such as a blemish or missing page, may be replicated in our edition. We do,
however, repair the vast majority of imperfections successfully; any imperfections that
remain are intentionally left to preserve the state of such historical works.

1 MONTH OF
FREE
READING

at
www.ForgottenBooks.com

By purchasing this book you are eligible for one month membership to ForgottenBooks.com, giving you unlimited access to our entire collection of over 700,000 titles via our web site and mobile apps.

To claim your free month visit:
www.forgottenbooks.com/free263291

English
Français
Deutsche
Italiano
Español
Português

www.forgottenbooks.com

Mythology Photography **Fiction**
Fishing Christianity **Art** Cooking
Essays Buddhism Freemasonry
Medicine **Biology** Music **Ancient**
Egypt Evolution Carpentry Physics
Dance Geology **Mathematics** Fitness
Shakespeare **Folklore** Yoga Marketing
Confidence Immortality Biographies
Poetry **Psychology** Witchcraft
Electronics Chemistry History **Law**
Accounting **Philosophy** Anthropology
Alchemy Drama Quantum Mechanics
Atheism Sexual Health **Ancient History**
Entrepreneurship Languages Sport
Paleontology Needlework Islam
Metaphysics Investment Archaeology
Parenting Statistics Criminology
Motivational

PROCEEDINGS

OF THE

LAKE SUPERIOR MINING INSTITUTE

SIXTH MEETING

FEBRUARY, 1900

VOL. VI

PUBLISHED BY THE INSTITUTE

OFFICERS OF THE

LAKE SUPERIOR INSTITUTE, 1900.

PRESIDENT.

Term expires.

GRAHAM POPE,	1901

VICE-PRESIDENTS.

O. C. DAVIDSON,	1901
T. F. COLE,	1901
J. H. MCLEAN,	1902
F. W. DENTON,	1902
M. M. DUNCAN,	1902

MANAGERS.

E. F. BROWN,	1901
EDWIN BALL,	1901
WALTER FITCH,	1901
G. H. ABEEL,	1902
J. B. COOPER,	1902

TREASURER.

A. J. YUNGBLUTH,	1901

SECRETARY.

F. W. SPERR,	1901

(The above officers constitute the council.)

HONORARY MEMBERS.

Raphael Pumpelly, Newport, R. I. T. B. Brooks, Newburg, N. Y.
N. H. Winchell, Minneapolis, Minn.

ACTIVE MEMBERS.

Abeel, G. H., Hurley, Wis.
Adams, John Q., Negaunee, Mich.
Alvar, Geo. A., Norway, Mich.
Anderson, G. A., Commonwealth, Wis.
Anderson, W. H., 1214 N. 3d St., Ishpeming, Mich.
Appleby, Wm. R., Minneapolis, Minn.
Bailey, C. E., Virginia, Minn.
Ball, Edwin, Soudan, St. Louis Co., Minn.
Baker, Fred J., Ishpeming, Mich.
Barr, H. A., Escanaba, Mich.
Bayley, W. S., Waterville, Me.
Bayliss, Willard, Ironwood, Mich.
Beattie, S. T., Florence, Wis.
Bennett, R. M., Minneapolis, Minn., New York Life Bldg.
Bennett, S. C., Crystal Falls, Mich.
Bjork, Arvid, Crystal Falls, Mich.
Blackwell. Frank, Ely, Minn.
Bohman, Chas. E., Iron Mountain, Mich.
Bond, William, Vulcan, Mich.
Boss, C. M.
Bradt, E. F., Iron Mountain, Mich.
Brady, Samuel, Rockland, Mich.
Breitung, Edward N., Marquette, Mich.
Brett, Henry, Calumet, Mich.
Brown, E. F., Iron Mountain, Mich.
Brown, Wm. C., Box 14, Brooklyn, N. Y.
Campbell, Duncan, Ishpeming, Mich.
Carbis, Frank, Iron Mountain, Mich.
Carbis, John, Iron Mountain, Mich.
Carey, John, 513 Rookery, Chicago, Ill.
Chadbourne, T. L., Houghton, Mich.
Channing, J. Parke, 34 Park Place, New York City.
Chynoweth, B. F., Rockland. Mich.
Clancy, James, Ishpeming, Mich.

Cole, T. F., Ironwood, Mich.
Cole, W. A., Ironwood, Mich.
Cole, William H., 117 East 3d St., Duluth, Minn.
Collick, Alfred, Ishpeming, Mich.
Conro, Albert, Milwaukee, Wis. New Insurance Bldg.
Collins, S. J., Milwaukee. Wis.
Cooper, James B., South Lake Linden, Mich.
Copeland, F., Vulcan, Mich.
Coventry, F. L., Commonwealth, Wis.
Cowling, Nicholas, Ely, Minn.
Crowell, Benedict, Cleveland, Ohio.
Cundy, James H., Quinnesec, Mich.
Curry, S. S., Ironwood, Mich.
Dalton, H. G., Cleveland, Ohio.
Davidson, O. C., Commonwealth, Wis.
Denton, F. W., Winona, Mich.
Dee, James R., Houghton, Mich.
Dengler, Theo., Atlantic Mine, Mich.
Dickinson, E. S., Norway, Mich.
Dubois. Howard W., 1527 N. 20th St., Philadelphia, Pa.
Duncan, John, Calumet, Mich.
Duncan, Murray M., Ishpeming, Mich.
Dyer, H. H., Bimabic, Minn.
Edwards, A. D., Atlantic Mine, Mich.
Ellard, H. F., Norway, Mich.
Elliott, Mark, Palmer, Mich.
Emmerton, F. A., Cleveland, Ohio.
Esselstyn. John N., Baker City, Oregon.
Fairbairn, C. T., Ishpeming, Mich.
Fisher, H. D., Florence, Wis.
Fisher, James, Houghton, Mich.
Fitch, Chas. H., Oak Park, Ill.
Fitch, Walter, Beacon, Mich.
Findley, A. I., 27 Vincent St., Cleveland, Ohio.
Flewelling, A. L., Crystal Falls, Mich.

Frangquist, August, Crystal Falls, Mich.

Gibbs, Geo., Baltimore Locomotive Works, Philadelphia, Pa.

Gilchrist, J. D., Iron Mountain, Mich.

Gibson, Thoborn, Amasa, Mich.

Goldsworthy, Martin, Iron Mountain, Mich.

Goodsell, B. W., 33 So. Canal St., Chicago, Ill.

Gordon, T. S., 171 Lake St., Chicago, Ill.

Goudie, Jas. H., Ironwood, Mich.

Grant, U. S., Minneapolis, Minn.

Greatsinger, J. L., Duluth, Minn.

Green, R. B., Two Harbors, Minn.

Haire, Norman W., Ironwood, Mich.

Hall, C. H., Ishpeming. Mich.

Hanna, L. C. Cleveland, Ohio.

Hardenburgh, L. M., Iron Mountain, Mich.

Harris, John L., Hancock, Mich.

Harris, S. B., Hancock, Mich.

Haselton, H. S., Sec'y Odanah Iron Co., Milwaukee, Wis.

Hauserman, Jacob, Commonwealth, Wis.

Hayden, George, Ishpeming, Mich.

Heath, Geo. L., South Lake Linden, Mich.

Hellberg, Gustaf A., Norway. Mich.

Hendrick, C. E., 424 So. Pine St., Ishpeming, Mich.

Hillyer, V. S., Iron Mountain. Mich.

Hodge, Chas. J., Houghton, Mich.

Holland, James, Iron Mountain, Mich.

Honnold, W. L., Room 15, 310 Pine St., San Francisco, Cal.

Hood, Ozni Porter, Houghton, Mich.

Hopkins, E. W., Commonwealth, Wis.

Horn, A. R., Milwaukee, Wis.

Houghton. Jacob, 51 E. Elizabeth St., Detroit, Mich.

Howe, John H., Duluth. Minn., care Marshall, Wells Hardware Co.

Hubbard, Lucius L., Houghton, Mich.

Hughes, C. W., Amasa. Mich.

Hulst, Nelson P., Milwaukee. Wis.

Hyde, Welcome, Appleton, Wis.

Johnston, Wm. H., Ishpeming, Mich.

Jones, B. W., Vulcan, Mich.

Jones, J. T., Iron Mountain, Mich.

Jopling, J. E., Ishpeming, Mich.

Judson, F. A., Trenton, N. J.

Karkeet, J. H., Iron Mountain, Mich.

Kaufman, Louis G., Marquette, Mich.

Kellerschon, J., Ironwood, Mich.

Kelly, William, Vulcan, Mich.

Kidwell, Edgar, 207 Front St., San Francisco, Cal.

Kirkpatrick, J. Clark, Palmer, Mich.

Knight, R. C., Commonwealth, Wis.

Koenig, Geo. A.. Houghton, Mich.

LaLonde, B. E., Ishpeming, Mich.

Lane, Alfred C., Lansing, Mich.

Lane, J. S., Akron, Ohio.

Larsson, Per, Roros, Norway.

Lasier, Fred G., Iron Mountain, Mich.

Lawrence, Chas. E., Amasa, Mich.

Lawton, C. A., Depere, Wis.

Leopold, Nathan F., Chicago, Ill., 402 Equitable Bldg., 108 Dearborn Street.

Lerch, Fred, Virginia, Minn.

Longyear, J. M., Marquette, Mich.

Luxmore, T. L., Iron Mountain, Mich.

Lyon, J. B., Norway. Mich.

Maas, Arthur E., Negaunee, Mich.

Manville, H. E., Milwaukee, Wis.

Martin, Thos. H., Ishpeming, Mich.

Mather, Wm. G., Cleveland. Ohio.

McComber, F. B., 19 S. Canal St., Chicago, Ill.

McDonald, D. B.

McGee, M. B., Crystal Falls, Mich.

McKee, W. E.

McLean, J. H., Ironwood, Mich.

McNair, F. W.. Houghton, Mich.

McNaughton, James, Iron Mountain, Mich.

McVeil, E. D., Crystal Falls. Mich.

Miller, L. B.. 39 Wade Building, Cleveland, Ohio.

Mills, F. P., Coulterville, Cal.

Mitchell, P., Hibbing, Minn.

Mixer, Charles T., Ishpeming, Mich.

Morgan, D. T., Republic, Mich.

Munroe, H. S., School of Mines, New York, N. Y.

Newett, Geo. A., Ishpeming, Mich.
Olcott, W. J., Duluth, Minn.
Oliver, H. W., Pittsburg, Pa.
Orrison, J., Iron Mountain, Mich.
Osborn, C. S., Sault Ste. Marie, Mich.
Palmer, E. V., Mass City, Mich.
Parker, Richard A., Cal. Exploration Co., San Francisco, Cal.
Parnall, S. A., Globe, Ariz.
Parnall, Wm. E., Calumet, Mich.
Parker, Wm. A., 509 Home Insurance Building, Chicago, Ill.
Parker, Richard A., San Francisco, Cal.
Paull, H. B., Globe, Ariz.
Peacock, Dan C., Commonwealth, Wis.
Pearce, Jas. H., Virginia, Minn.
Pearse, Frank E.
Pengilly, John, Ely, Minn.
Pitkin, S. H., Akron, O.
Pope, Graham, Houghton, Mich.
Prescott, F. M., Oregon St., Milwaukee, Wis.
Rattle, W. J., 807 Western Reserve Building, Cleveland, O.
Redfern, J. A., Hibbing, Minn.
Roberts, E. S., Iron River, Mich.
Robinson, A. W., South Milwaukee, Wis.
Roberts, R., Negaunee, Mich.
Roscorla, N. B., Ironwood, Mich.
Rose, R. S.
Rothwell, H. G., i. c. o. E. N. Breitung, Marquette, Mich.
Rough, James H., Ishpeming, Mich.
Rundle, A. J., Iron Mountain, Mich.
Ryan, Edward, Hancock, Mich.
Scadden, Frank, Crystal Falls, Mich.
Scott, Dunbar D., Houghton, Mich.
Seaman, A. E., Houghton, Mich.
Shephard, Amos, Iron Mountain, Mich.
Simpson, C. S., Florence, Wis.

Sjögren, S. A., Hjalmar, Upsala, Sweden.
Smyth, H. L., Cambridge, Mass.
Sperr, F. W., Houghton, Mich.
Stabler, O. F., Ironwood, Mich.
Stackhouse, Powell, Wallingford, Pa.
Stanton, F. McM., Atlantic Mine, Mich.
Stanton, John, 11 and 13 William St., New York City.
Stanton, John R., 11 and 13 William St., New York City.
Stephens, James, Ishpeming, Mich.
Streeter, Albert T., Calumet, Mich.
Sturtevant, H. B., Ely, Minn.
Sutherland, D. E., Ironwood, Mich.
Sweeney, E. F., Virginia, Minn.
Swift, Geo. D., Duluth, Minn.
Thompson, A. W., Vulcan, Mich.
Thompson, Jas. R., Ironwood, Mich.
Tregouning, Jos., Ishpeming, Mich.
Trepanier, Henry, Iron Mountain, Mich.
Trestrail, William C., Iron Mountain, Mich.
Truscott, Henry, Loretto, Mich.
Tweedy, R. B., Milwaukee, Wis.
Van Dyke, Wm. D., Milwaukee, Wis.
Van Dyke, John H., care Pewabic Co., Milwaukee, Wis.
Van Hise, C. R., Madison, Wis.
Van Mater, J. A., Elvins, St. Francois Co., Mo.
Vickers, J. M., Ishpeming, Mich.
Wadsworth, M. E., P. O. Box 296, Chicago, Ill.
Wenstrom, Olof, Sulitjelma, Norway.
White, Peter, Marquette, Mich
Whiting, S. B., Calumet, Mich.
Winchell, Horace V., Butte. Mont.
Woodworth, George L., Iron River, Mich.
Yungbluth, A. J., Ishpeming, Mich.

DECEASED MEMBERS.

Bawden, John T.
Dickinson, Wm. E.

Holley, S. H.

CONTENTS.

RULES OF THE INSTITUTE.

I.

OBJECTS.

The objects of the Lake Superior Mining Institute are, to promote the arts and sciences connected with the economical production of the useful minerals and metals in the Lake Superior region, and the welfare of those employed in these industries, by means of meetings for social intercourse, by excursions, and by the reading and discussion of practical and professional papers, and to circulate, by means of publications among its members, the information thus obtained.

II.

MEMBERSHIP.

Any person interested in the objects of the Institute is eligible for membership.

Honorary members, not exceeding ten in number, may be admitted to all the privileges of regular members except to vote. They must be persons eminent in mining or sciences relating thereto.

III.

ELECTION OF MEMBERS.

Each person desirous of becoming a member shall be proposed by at least three members, approved by the Council, and elected by ballot at a regular meeting (or by ballot at any time conducted through the mail, as the Council may prescribe) upon receiving three-fourths of the votes cast. He shall become a member on the payment of his first annual dues, within ninety days of the notification of his election.

Each person proposed as an honorary member shall be recommended by at least ten members, approved by the Council, and elected by ballot at a regular meeting (or by ballot at any time, conducted through the mail, as the Council may prescribe), on receiving nine-tenths of the votes cast.

IV.

WITHDRAWAL FROM MEMBERSHIP.

Upon the recommendation of the Council, any member may be stricken from the list and denied the privilege of membership, by the vote of three-fourths of the members present at any regular meeting, due notice having been mailed in writing by the secretary to him.

V.

DUES.

The dues of members shall be five dollars, payable upon their election, and five dollars per annum thereafter, payable in advance at or before the annual meeting. Honorary members shall not be liable to dues. Any member not in arrears may become a life member by the payment of fifty dollars at one time, and shall not be liable thereafter to annual dues. Any member in arrears may, at the discretion of the council, be deprived of the receipt of publications or be stricken from the list of members when in arrears six months: PROVIDED, That he may be restored to membership by the council on the payment of all arrears, or by re-election after an interval of three years.

VI.

OFFICERS.

There shall be a president, five vice-presidents, five managers, a secretary and a treasurer, and these officers shall constitute the Council.

VII.

TERM OF OFFICE.

The president, secretary and treasurer shall be elected for one year, and the vice-presidents and managers for two years, except that at the first election two vice-presidents and three managers shall be elected for only one year. No president, vice-president or manager shall be eligible for immediate re-election to the same office at the expiration of the term for which he was elected. The term of office shall continue until the adjournment of the meeting at which their successors are elected.

Vacancies in the Council, whether by death, resignation, or the failure for one year to attend the Council meetings, or to perform the duties of the office, shall be filled by the appointment of the Council, and any person so appointed shall hold office for the remainder of the term for which his predecessor was elected or appointed: PROVIDED, That such appointment shall not render him ineligible at the next election.

VIII.

DUTIES OF OFFICERS.

All the affairs of the Institute shall be managed by the Council, except the selection of the place of holding regular meetings.

The duties of all officers shall be such as usually pertain to their offices, or may be delegated to them by the Council.

The Council may in its discretion require bonds to be given by the treasurer, and may allow the secretary such compensation for his services as they deem proper.

At each annual meeting the Council shall make a report of proceedings to the Institute, together with a financial statement.

Five members of the Council shall constitute a quorum; but the Council may appoint an executive committee, or business may be transacted at a regularly called meeting of the Council, at which less than a quorum is present, subject to the approval of a majority of the Council, subsequently given in writing to the secretary and recorded by him with the minutes.

There shall be a meeting of the Council at every regular meeting of the Institute and at such other times as they determine.

IX.

ELECTION OF OFFICERS.

Any five members, not in arrears, may nominate and present to the secretary over their signatures, at least thirty days before the annual meeting, the names of such candidates as they may select for offices falling under the rules. The Council, or a committee thereof duly authorized for the purpose, may also make similar nominations. The assent of the nominees shall have been secured in all cases.

No less than two weeks prior to the annual meeting, the secretary shall mail to all members not in arrears a list of all nominations made, and the number of officers to be voted for in the form of a letter ballot. Each member may vote either by striking from or adding to the names upon the list, leaving names not exceeding in number the officers to be elected, or by preparing a new list, signing the ballot with his name, and either mailing it to the secretary, or presenting it in person at the annual meeting.

In case nominations are not made thirty days prior to the date of the annual meeting for all the offices becoming vacant under the rules, nominations for such offices may be made at the said meeting by five members not in arrears, and an election held by a written or printed ballot.

The ballots in either case shall be received and examined by three tellers appointed at the annual meeting by the presiding officer; and the persons who shall have received the greatest number of votes for the several offices shall be declared elected. The ballots shall be destroyed, and a list of the elected officers, certified by the tellers, shall be preserved by the secretary.

X.

MEETINGS.

The annual meeting of the Institute shall be held at such time as may be designated by the Council. The Institute may at a regular meeting select the place for holding the next regular meeting. If no place is selected by the Institute it shall be done by the Council.

Special meetings may be called whenever the Council may see fit; and the secretary shall call a special meeting at the written request of twenty or more members. No other business shall be transacted at a special meeting than that for which it was called.

Notices of all meetings shall be mailed to all members, at least thirty days in advance, with a statement of the business to be transacted, papers to be read, topics for discussion and excursions proposed.

No vote shall be taken at any meeting on any question not pertaining to the business of conducting the Institute.

Every question that shall properly come before any meeting of the Institute shall be decided, unless otherwise provided for in these rules, by the votes of a majority of the members then present.

Any member may introduce a stranger to any regular meeting; but the latter shall not take part in the proceedings without the consent of the meeting.

XI.

PAPERS AND PUBLICATIONS.

Any member may read a paper at any regular meeting of the Institute, provided the same shall have been submitted to and approved by the Council, or a committee duly authorized by it for that purpose prior to such meeting. All papers shall become the property of the Institute on their acceptance, and, with the discussion thereon, shall subsequently be published for distribution. The number, form and distribution of all publications shall be under the control of the Council.

The Institute is not, as a body, responsible for the statements of facts or opinion advanced in papers or discussions at its meetings, and it is understood, that papers and discussions should not include personalities, or matters relating to politics, or purely to trade.

XII.

AMENDMENTS.

These rules may be amended by a two-thirds vote taken by letter ballot in the same manner as is provided for the election of officers by letter ballot: PROVIDED, That written notice of the proposed amendment shall have been given at a previous meeting.

MINUTES OF THE SIXTH ANNUAL MEETING

OF THE

LAKE SUPERIOR MINING INSTITUTE

HELD ON THE MENOMINEE RANGE

WITH IRON MOUNTAIN AS THE CENTER, FEB. 6, 7, AND 8, 1900.

Tuesday, February 6, at 9 A. M., the members and guests of the Institute, to the number of about 100, took the special train furnished for the occasion by the Chicago & Northwestern Railway Company for a trip to the eastern end of the range.

The first stop was made at Quinnesec for a visit to the Cundy Mine, owned and operated by the Illinois Steel Company. The ore is in a different geological horizon from the other iron ores which are mined on this range; and the ore is very hard. The surface plant is conveniently arranged for handling the product.

The next point visited was the little town of Niagara, situated about one mile south of Quinnesec on the lower falls of the Menominee river. Here, on the Wisconsin side of the river, the Kimberly & Clark Company are just completing a large pulp and paper mill. When all the buildings are finished the ground space covered will be 956 feet long by from 75 to 100 feet wide.

The wood goes from the saw mill to the barking machine, where the bark is removed, then to the grinding mill in which there are sixteen machines for the reduction of the wood, and then into vats where it is made into pulp.

The pulp is now being shipped to other mills for finishing; but the machinery is in place and will be ready to treat the material in this mill by the first of next April. The finishing rolls are brass covered to avoid rust. The drying rolls are heated by steam, 24 in number, each weighing 7,500 pounds.

The mill will turn out daily 100 tons of manila and wall

paper. Hemlock is used for the manufacture of wall paper. and spruce and pine are used for lower grades of paper.

There is a sulphite mill which furnishes the chemical for the digestors in which the hemlock is treated. There are two digestors each 48 feet high by 14 feet in diameter, made of steel sheets 1⅛ inch thick. A little blue vitriol is used in the digestors, and the wood is drawn off once every day.

·The next stop was made at Norway, where the Aragon mine was the principal point of interest to the members. In this mine the ninth level is now being opened 100 feet below the eighth level. with drifts large enough for the operation of a pneumatic haulage plant which is now being successfully employed in the eighth level cross-cut from the ore body to the shaft, a distance of 700 feet. The drifts of the eighth level are not large enough to accommodate the locomotive which requires a height of five feet and a width of seven feet on a straight run, and a greater width on the curves. Mules are now employed for gathering the ore in the eighth level drifts to the place where the trains are made up for the locomotive.

The compressed air is furnished to the locomotive by a Norwalk high pressure triple stage compressor, single steam cylinder, the gauge showing 600 pounds in the high pressure cylinder.

About 1000 gallons of water per minute is pumped from the bottom of the mine by a triple expansion Worthington pump. About 500 gallons per minute of surface water is pumped from the 225-foot level.

The steel shaft-house at the Harrison shaft attracted much attention, and elicited the general opinion that it is an improvement on the wooden structure. It is claimed that the steel shaft-house can be built stronger and will last longer than the wooden shaft-house; and it will not burn.

From the Aragon mine the party went to the Penn Iron Company's mines. The first of these mines visited is at West Vulcan. Here the party was most particularly interested in the surface tramming plant. This consists of an 8-wheel car running at high speed and on reverse curve on a trestle from the shaft-house to the stock pile, and so arranged that it delivers its load at different points, making separate piles of the different grades of ore. It is a side dumping car with a detaching apparatus which is set by the surface lander according to the signal which he receives from below of a certain number of bells indicating a certain grade of ore.

This mine was opened on what is known as the south deposit, and exploited to the depth of 1000 feet, when the north deposit was discovered in the lower levels. The north deposit was then worked from the bottom upward by a successful method of rooming and filling and taking out pillars by caving. The mine makes a large quantity of water which is handled by a triple expansion Worthington pump. A compound Worthington pump and a Cornish pump are held in reserve.

The East Vulcan mine, owned by the same company, was the next point visited. This mine has been idle several years. It is now being unwatered by a specially contrived bailing arrangement which attracted the notice of the members, and has been made the subject of a paper presented by Mr. Kelly.

From the East Vulcan the party went to the Verona mine, which was formerly the Southeast Vulcan. A new machinery plant is being installed, and the shaft is nearly down to the 4th level. The ore body extends under the Sturgeon river near its junction with the Pine river; and because of this the ore cannot be mined by caving, but large pillars are left to protect the surface. From here the party returned to Iron Mountain for the evening session.

At 8 o'clock P. M. the meeting at the Court House was called to order by President William Kelly, who delivered the presidential address.

The next paper was read by Mr. L. M. Hardenburgh of the Pewabic mine, on the subject of crushing and concentrating iron ore at the works at Keel Ridge.

Mr. A. W. Thompson, electrician of the Penn Iron Mining Company, presented a paper on the system of bells established at the West Vulcan mine. This paper brought out quite a general discussion of systems of bell signals.

Mr. James MacNaughton, general manager of the Chapin Mining Company, read a paper on mine dams and the unwatering of shafts, drifts, etc., in the event of sudden influx of water.

The remainder of this evening's session was devoted to routine business. Messrs. Van Dyke, Newett, Abeel, Davidson and Denton were appointed a nominating committee to present at the next session names for officers for the ensuing year. Messrs. Morgan, Beattie and Bradt were appointed an auditing committee. Twenty-four names were elected to membership, after which the meeting adjourned.

Wednesday, February 7. At 9 A. M. the special train left Iron Mountain to take the party to the mines on the western end of the range.

The first property visited was the Dober mine at Stambaugh, which is one of a group of four mines operated by the Oliver Mining Company in the Iron River valley. Here is a large body of non-bessemer ore over which Iron river flows, and a swamp lies over a portion of the deposit. Careful calculations have been made from all available data, of the probable high water flow from the drainage system of the river above the mine; and the river will be diverted through a new channel having a capacity of 2,000 cubic feet per second with a fall of $2\frac{1}{2}$ feet, in a distance of about 2.200 feet. A dyke will be built around on three sides of the property, on the upper side to divert the stream, on the river side to prevent inflow, and on the lower side to prevent backflow. A trench through the muck of the swamp will be cut out to the underlying clay upon which the dyke will be built.

The Riverton mine, which is operated by the same company, was next visited. This also underlies the river which is projected to be carried over the property in a flume one-fourth of a mile long.

The ore body and enclosing walls are firm and no timber is used in mining. The mine was idle for several years, but no falling or caving of ore or rock occurred.

The next visit was made to the Mansfield mine, where the awful tragedy of August 23, 1893, occurred, when the Michigamme river broke through and flooded the mine, drowning 28 miners. It had been attempted to support the surface with very heavy timbering at great cost. But the best of timbering was ineffectual; and since the catastrophe the river has been diverted at less expense, it is said, than it cost to put in the extra timbering.

The remainder of the day was occupied at Crystal Falls, the principal center of this end of the range.

The Crystal Falls mine, the Great Western, the Lincoln, and the Monitor are operated by Messrs. Corrigan, McKinney & Co. The points of interest were the methods and operations for the renewal and revival of these properties, using the old machinery for all the work.

The Bristol mine is operated by the Commonwealth Iron Mining Company. This is a large deposit of low grade ore worked by the open pit and milling methods. The ore is hard and breaks in large pieces, and a Gates crusher will be installed to break the pieces of ore into marketable size. It was now late in the day and the trip to Florence and Commonwealth had to be abandoned.

At 8 o'clock P. M. the second evening session was called to order by President Kelly in the Court House of Iron County at Crystal Falls.

The first paper presented was by Mr. C. H. Fitch, on "Economy in the Manufacture of Mining Machinery." In the absence of the author the paper was read by the acting secretary, President F. W. McNair, of the Michigan College of Mines. An interesting discussion of the subject followed the reading of the paper.

A paper on the "Method of Mining at the Badger Mine" was presented by Mr. O. C. Davidson, superintendent of the Commonwealth Mining Company, and of the Antoine Ore Company.

The last paper on the program was presented by Mr. William Kelly on "Balancing Bailers."

Mr. Kelly exhibited a model of the contrivance, and much interest was taken in the subject by the mining men present.

The following routine business was transacted:

The names of 29 applicants were elected to membership, making a total of 53 new members.

An invitation from the Rock Lake Mining Company of Canada was extended through Mr. B. W. Goodsell to all members of the institute to visit their properties. This was placed on file with the thanks of the Institute.

Invitations from the Citizens' Business League, Mayor Rose and the Journal of Milwaukee, to have the next meeting of the Institute held in that city, were read and placed on file.

Mr. J. B. Cooper, presented a resolution thanking the Milwaukee people for the invitations, together with a motion that the next place of meeting be left to the decision of the executive council.

Mr. O. C. Davidson presented a resolution thanking the officials of the Chicago & Northwestern Railway Company, particularly Superintendent W. B. Linsley of this division, for the many courtesies extended.

Mr. T. F. Cole moved a vote of thanks in behalf of the membership to President Kelly for the efficient manner in which he discharged the duties of his office during his term.

2

The following officers were recommended by the nominating committee and elected by unanimous vote:

President, Graham Pope, of Houghton; vice presidents, O. C. Davidson, T. F. Cole, P. H. McLean, F. W. Denton, and M. M. Duncan; managers, E. F. Brown, Edwin Ball, Walter Fitch, George H. Abeel and J. B. Cooper; treasurer, A. J. Yungbluth; secretary, F. W. Sperr.

Mr. Pope was escorted to the chair and with his usual felicity expressed his appreciation of the honor conferred upon him in the following words:

Mr. President, Gentlemen of the Institute:

I accept the office to which you have elected me. I accept it with a grateful appreciation of a high honor conferred upon me. For surely it is a high honor, and one which any man may be justly proud to receive from this Institute, composed. as it is, of men eminent in their several positions, who have done so much for the advancement of the mining industry, and whether as bosses, captains, engineers, superintendents or other officers, have always had a lively sense of the great responsibilities resting upon them, and upon whom has depended the financial success of those who placed them there.

I congratulate the Institute upon its great success, its steady growth, its promise of future usefulness, and the public appreciation of its value as is shown by the telegrams of invitation which have come to us from Milwaukee. Gentlemen, I will perform the duties of my office with fidelity and zeal, and with such success as may be granted to me. I thank you heartily.

After the adjournment of the meeting, the party returned to Iron Mountain, arriving there about midnight.

Thursday, February 8. The forenoon was occupied in a trip of inspection to the Pewabic mine. The chief point of interest here is the block caving system of mining, described by Mr. Brown in Vol. V of the Proceedings.

In the afternoon the Chapin mine was visited. This is the largest mine upon the Menominee Range and presents many interesting features. On the surface the points of special interest were the vertical skip dumping arrangement; the car for automatically distributing the ore upon the stock pile according to the grade of ore; the flat rope hoisting engine; the mill for framing mine timbers; and the use of compressed air brought from the compressor plant at Quinnesec Falls, for operating the hoisting engines and all other machinery.

Underground the points of special interest are the large Riedler pumps, the dams holding back large quantities of water under high pressure, the chain haulage plant, the sub-stoping system of mining, and the pockets for loading the skips.

In the evening the meeting of the Institute closed with a banquet at the Commercial hotel. After the excellent menu had been disposed of, short speeches were made by Hon. Chase S. Osborn of Sault Ste. Marie, Supt. J. B. Cooper of the Calumet and Helca Smelting Works, Mr. J. Parke Channing of New York City, and Attorney Geo. W. Hayden of Ishpeming, with President Pope acting as toastmaster.

Then, under a suspension of the rules, four more applicants were elected to membership. The meeting adjourned at 2 A. M. Friday.

CANDIDATES APPROVED BY THE COUNCIL AND ELECTED TO ACTIVE MEMBERSHIP.

G. A. Anderson.
Charles E. van Barneveld.
H. A. Barr.
S. C. Bennett.
Arvid Bjork.
Duncan Campbell.
Frank Carbis.
John Carbis.
John Carey.
Wm. H. Cole.
James H. Cundy.
H. G. Dalton.
E. S. Dickinson.
S. R. Elliott.
H. D. Fisher.
A. L. Flewelling.
August Frangquist.
Norman W. Haire.
L. M. Hardenburgh.
Jacob Harper.
H. S. Haselton.
V. S. Hillyer.
O. P. Hood.
J. H. Howe.
Welcome Hyde.
Geo. L. Jackson.
Frank A. Janson.
Anton Johnson.

B. W. Jones.
F. G. Lasier.
N. F. Leopold.
John Luxmore.
A. E. Maas.
F. B. McComber.
D. B. McDonald.
W. E. McKee.
F. W. McNair.
F. D. McNeil.
Gordon Murray.
W. H. O'Leary.
Thomas Pascoe.
D. C. Peacock.
James H. Pearce.
Sam'l Perkins.
A. J. Rundle.
Chas. B. Schaefer.
Amos Shephard.
James Stephens.
John T. Spencer.
John R. Stanton.
L. T. Sterling.
A. T. Streeter.
A. W. Thompson.
Henry Trepanier.
Wm. C. Trestrail.
Henry Truscott.

Geo. L. Woodworth.

FINANCIAL STATEMENT.

RECEIPTS.

Dues from members	$745	18
Sales of proceedings	18	50
Extra copies to authors	5	00
Interest on deposits	10	16
	$778	84
Balance from last report	100	90
		$879 74

EXPENDITURES.

Printing	$377	40
Transportation	20	92
Sundry office expenses	57	18
Furniture	13	25
Secretary's salary	100	00
		568 75
Balance on hand		$310 99

ADDRESS OF PRESIDENT WILLIAM KELLY.

The present condition of the mining business is in striking contrast to that prevailing for the six years prior to the summer of 1899. We rejoice in a glorious prosperity to which we feel entitled in return for most exacting sacrifices, and which we fully appreciate in comparing our lot with what we have been through. It is a real prosperity because it is shared by all. We may not hope that hard times will never come again, but the business skies are bright and we have reason to expect that the period of prosperity will linger through a long cycle of years. Because of present contentment we can look back without bitterness and before our memory becomes clouded with new figures, new pictures, it seems proper to recall the trying conditions now so happily ended and record some of the lessons which experience has taught.

The two principal industries in which the members of our Institute are engaged are both subject to great changes in demand and price. A fall in the price of copper may not occur at the same time as that in iron ore, and some of us may have thought that our friends in the copper country do not know what hard times are, and yet the average price of copper in '94 was less than half that of 1880. The reports of the Commissioners of Mineral Statistics of Michigan record the average price of copper in New York for each month and year and I have placed on the chart the average yearly price since 1860, though as prices were quoted in paper money values prior to 1879, they are hardly comparable with those since. After the good times of '79 and '80 there was a gradual but steady fall in the average yearly price and the failure of the French corner in copper, together with the general depression in business which followed the Baring failure, brought down the price of copper until the average for 1886 was 11.00 cents, 9.18 cents less than that prevailing six years before,—a reduction of over 45 per cent. In 1890 the average price was 15¾ cents, but it fell 6 3-16 cents or nearly 40 per cent in four years, when in 1894 the average price of Lake copper for the year was 9 9-16 cents. These depressions indicate the pressure placed upon the copper mines to reduce the cost of production.

Unlike lake copper, iron ore is variable in composition and

value. Data is wanting from which can be computed the average price of all ores, and the prices placed on the chart are those quoted in reports of Commissioners of Mineral Statistics for Michigan as Standard Bessemer iron ore at lower lake ports. Where a variation in the price is noted as since 1892 the highest price given has been used. We, of the younger generation of Lake Superior miners, have often heard glowing stories of the golden period of twelve-dollar ore, but that price is recorded for one year only, a year more memorable for the Jay Cook failure, the panic year of 1873. After the resumption of specie payments in '79 the iron business had its record boom. For three years the price was high. Then a lower plane was accepted and for the succeeding ten years, beginning with '83 the price from year to year varied less than before.

Of the ten years the price of

Five years, viz.,	'85, '86, '87, '88, '89 and '92 was			$5.50
One year	"	'84	"	5.75
" "	"	'91	"	6.00
" "	"	'83	"	6.25
" "	"	'90	'	6.50
" "	"	'87	'	7.25

The average of the ten years was $5.92.

Passing over '93, which was disturbed by the silver panic, the next six years show as follows:

1894	$2.75
1895	3.50
1896	4.50
1897	3.18
1898	3.64
1899	3.65

giving an average of $3.54, a reduction of 40 per cent.

The figures given for the later years cover ores now considered special, the price of the fictitious ore of fixed composition which has been taken as the basis for valuating Bessemer ore for the last three years being respectively 53, 89 and 70 cents less. Were it possible, however, to compute the average price of all ores, the reduction would be equally striking.

How the change in values was met by the mines, it is the purpose of this paper to consider.

At the outset, credit must be given for the share of the reduction borne by the transportation interests. But, how it was possible for vessels to carry ore from Duluth to Lake Erie ports for 60 cents a ton, and how the vessels have developed in size and equipment, is another story. And this is not the place to consider in detail the modernizing of the Lake Superior railroads. The importance of one tendency of industrialism,—the increase in the size of the units handled,—both parts of the transportation system clearly demonstrate.

The fall in price necessarily bore down first on profits. But they were quite insufficient for such a strain and the pressure of reduction had to be met principally in the cost.

The period under consideration is not marked by any notable invention. Nitro-glycerine and the power drill had long been commonplaces. Nor can the results be attributed to the general introduction of previously unknown methods or the extensive application of untried forces. It is impossible to point out one or even any few causes that explain the reduction of costs. At every mine the situation was met in a different way. The variety of the solutions, the ingenuity of their conception and the courage of their adoption make the subject of the reduction of costs of mining ore on the older ranges of Lake Superior a fitting topic to present to this Institute.

Viewed from the standpoint of an accountant, the cost of ore is found by multiplying the amount of labor by the rates of wages, adding the supplies consumed multiplied by the price, adding again certain charges such as royalty, taxes, insurance, etc., and dividing the sum by the output. It seems proper to present the subject in this manner, because it emphasizes the importance of the output. Many elements make up the expenditure, but the divisor consists of one term only. The surest way to reduce the cost is to increase the output, while every element of the expenditure may be increased if it enables a corresponding or greater increase in the output.

Few who have not been through it can appreciate the agony of heart which seized us when in the fall of '93 and the opening months of '94 we realized what great reductions must be made in cost to insure a continuance of operations and were confronted with the appalling condition that at the same time the output must be reduced enormously. It seemed as if, like the children of Israel in Egypt, we were asked to make bricks without straw. As dead-work and the harder stopes were discontinued, it was pathetic to realize that even more men had to be turned adrift than the proportional reduction in output indicated.

It is much to be regretted that no record of the yearly out-

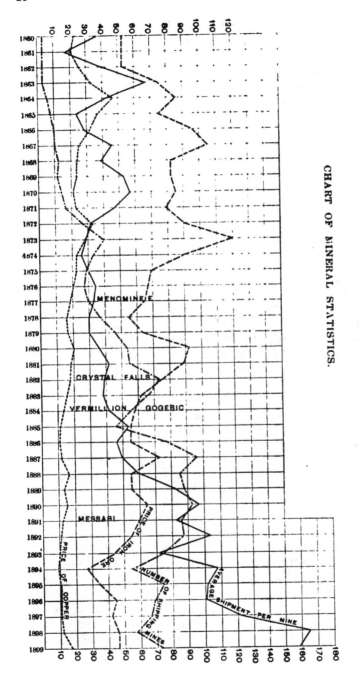

CHART OF MINERAL STATISTICS.

put of the various mines is available. It would show unmistakably to what extent the reduction of cost has been due to increased production. In some years the shipments are considerably less than the output and in others correspondingly exceed it, but a general comparison of the shipments of the different periods will throw some light on the tendency of recent efforts and I have therefore placed on the chart the number of "old range" iron mines that shipped each year, and the average shipment for that number. In the reports of the Commissioners of Mineral Statistics and the tables issued by the Iron Trade Review there are some discrepancies, but the occasional errors will not interfere with a few general deductions. Throughout, Norrie and East Norrie are counted as one, the Regent group and the Cleveland group each as one. Other mines that are grouped in some of the reports have been separated throughout. The number of shipping mines is not material to the present discussion, but when the figures were plotted a relation between the course of prices and the number of mines seemed to appear and may be of interest.

The Jackson, Cleveland and Lake Superior mines were the only shippers from 1860 to 1863, when the number began to increase until in 1873 there were 40 shipping mines. The hard times that followed reduced the number to 28 in '76. In '77 the Menominee Range was opened and under the stimulus of better prices the number of shippers rose to 74 in 1882, when the Crystal Falls and Iron River mines began to produce. For the next three years prices fell and the number of shipping mines receded to 46. The first ore from the Vermillion and Gogebic ranges came in '84. The great activity on the latter range swelled the number rapidly and the largest number of shipping mines is recorded for 1887, when there were 96 on the list. The same year shows a maximum point in the price line. Both lines show a minimum in '88, a maximum in '90, and a minimum in '94. In that year there were only 55 shipping mines. Thereafter the courses of the two lines bear only a general resemblance.

From '72 to '85 the yearly average shipments were under 40,000 tons. Eighteen hundred eighty-seven was the hey-day of the small mine. A 50,000 tons mine was then above the average. In 1892 the average rose to over 100.000 tons for the first time. From 1892 to 1896 may be called the 100,000 tons period, the average for

3

1892 being .... 104,000 tons.
1893, under the influence of the panic, being... 73,000 tons.
1894 being .... 108,000 tons.
1895 being .... 102,000 tons.
1896 being .... 101,000 tons.

A part of the shipments for '94, '95 and '96 was ore carried over from previous years and during that time the producing power of the mines was restrained. In 1897 the average of shipments rose to 120,000 tons, in 1898 to 165,000 tons and in 1899 it was 159,000 tons. From this it appears that larger outputs have assisted to reduce costs principally during the last three years.

The deposits of ore from which the tremendous shipments of recent years have been made, with perhaps one or two exceptions, were known and more or less exploited before the panic. The difficulty in marketing ore held back its extraction, and on the other hand, the possibilities of development of the older ranges, together with the unfolding of the wealth of the Mesabi, were potent factors in bearing prices and retarding effort.

By far the largest part of the expenditures in mining operations is for labor. During the hard times every effort was put forth to reduce the amount of labor per ton. When the panic broke, the mines generally, were in a fair state of development. The only question being as to whether they should run at all, prospecting was abandoned and dead-work stopped. This was a great help to the cost for a couple of years, but after a while mining caught up to the developments and dead-work had to be resumed. Stopping the dead-work could be only a temporary expedient in the reduction of costs. It served its purpose and tided over the crisis.

In many cases a permanent reduction in the amount of labor has been found in a change in the system of mining or in a change of details in its application and it is probable that the necessities of the period have in some measure helped to effect the change. Where conditions made it possible to adopt a caving or settling system, the saving was often considerable, both in work and in timber and there are a number of instances where this may be considered as the greatest gain that has been made, but it cannot be considered as a general explanation of the reduction in costs, because in a great many mines the conditions did not permit the adoption of such a method. Such mines sought other ways to reduce. Some were able to make improvements in their mining system. Thus in rooming in timbers, it was possible at some places to catch more ore in chutes than had formerly been the practice, and thereby save in shoveling.

The substitution of other power for hand power has been hastened. The field which afforded the greatest opportunity for development was that of transporting the ore from stopes to stockpiles and pockets. Before the panic three rope haulage plants had been established underground, though two of them were afterwards given up. The largest deposits under Lake Angeline were a particularly favorable field for mechanical haulage and electric systems were adopted before the panic. Subsequently at a number of other mines, mules were successfully introduced and helped materially to reduce the cost. Later a chain haulage plant was installed at the Chapin and the latest appliance to bid for honors is the pneumatic locomotive at the Aragon. A reduction in the cost of tramming permits of a further important saving by increasing the economical distance between shafts. Fewer shafts need be sunk and greater quantities of ore are made tributary to each. Some changes, too, may be noted in the method of hoisting. In many of the deeper mines, where the hoisting of men is an important matter. cages were early adopted and the practice with them was formerly quite general of raising the mine car to the surface and carrying the ore in it from stope to pocket. When different grades were being separated this plan seemed to afford less danger of mistake or mixing. But many of these mines, while retaining cages to transport men and timber, now hoist the ore in skips. The danger of mixing was a bug-a-boo and the saving has been considerable. Hoisting in skips leads naturally to some mechanical arrangement for conveying the ore from shaft to pocket or trestle and the last few years have seen several commendable methods put into operation.

The size of the units handled has increased rather slowly. In the copper mines, owing to their greater depth the element of time is a serious limit to the production of a shaft and greater units are handled than are usual in the iron mines. The heaviest load of ore at the iron mines is five tons and the number of mines at which three tons are hoisted is very small. The usual load is less than two tons. In many cases the pay load is but a small fraction of the capacity of the hoisting plant. The last few years have undoubtedly seen progress, but there still remains room for considerable improvement at many mines, both in the size of the cars and in the loads for the hoist.

Steam shovels have been used about the mines for a number of years, but at the beginning of the period under consideration, there were still many mines which relied on hand labor for loading the stockpiles. Every spring appeared stout, hardy men who applied for stockpile work. But they came in fewer numbers until now the stockpile man has ceased to exist. In some cases the steam shovel may have driven him away but in many others the shovel filled a vacancy. It may be said in general terms, that the steam shovel has multiplied the efficiency of labor in the neighborhood of six fold and reduced the cost of loading one-half.

We cannot note any extension of the use of machinery for the framing of timber, for the reverse is undoubtedly the case. This is due partly to the use of smaller or less timber and partly to the fact that the framing mills so far tried have not been satisfactory. It is, however, a proper field for the use of machinery.

Under the necessity for retrenchment, the duties of every position were carefully scrutinized and readjusted. Thus, a skip tender was dispensed with at one level by putting in a speaking tube to the next, or a skip tender was entrusted with the care of the pump, or a dry-house man required to look after the oils or an oil man to make wedges. Those who had held the abolished places were not necessarily dismissed, but when possible, were added to the forces more actively engaged in increasing production.

The question as to how much attention should be paid to appearances is always with us, but hard times naturally magnify the utilitarian side of the case.

We come now to the second factor in the general formula which we have used to represent expenditure,—the price of labor. Upon the reduction of forces in 1893, wages were reduced; for the most of 1894 they were 25 per cent lower than they had been in 1891 and 1892, and the efficiency of labor was excellent. Though wages rose slightly in 1895, it was a year of dissatisfaction. In the early part of 1896 wages were raised to the 1892 level in expectation of a good year, but, with the political outlook threatening the standard of values, the hopes were not realized, and wages and prices fell again, to gradually rise once more in '98 and '99 with the advent of prosperity, until now they are generally higher than this generation of workers has known.

We have learned that the efficiency of labor does not depend upon the absolute amount of the wage, but upon its fairness in consideration of the condition of business. The fair and equitable rate of wages may be sometimes high and sometimes low. If both parties concede the fairness of the rate, labor is efficient; if the basis is unfair or thought to be unfair there must be loss or friction.

The amount of supplies consumed received most careful attention. Fuel, timber and powder are usually the principal items in the list, though some mines with water power require little fuel, others have such conditions of rock and ore as to need little timber and some require but little powder. We have heretofore considered under the head of labor saving, some economies that have been effected by the greater use of mechanical power. At the same time there has been a great improvement in machinery which has reduced the fuel bill. As the depth of mines increases and the flow of water enlarges, pumping, necessarily continuous, takes a large amount of power. When it becomes necessary to lift 1,800,000 tons of water a year 1000 feet, 2,000,000 tons 800 feet, 1,300,000 tons 1,330 feet, 2,200,000 tons 560 feet or 2,200,000 tons 750 feet, figures furnished by four mines on the Menominee Range, the principal question is what machine is the most economical and we have seen installed several triple expansion direct acting pumps and one compound fly wheel pump, giving duties of over 90 million foot pounds. High class compressors with compound steam and air cylinders and improved valve motions have been coming into use as new ones or larger ones are needed. Too often compressors are continued in use that consume an excess of fuel, enough to pay for a good compressor in a very few years. The palm is easily conceded to the copper mines for hoisting engines and even on the iron ranges they are, as a rule, of larger and better types than the other machinery. The object of mining being to raise ore, hoisting engines naturally receive early attention, so that it is not surprising that the number of new installations of late years has been comparatively small. Several water tube boilers of different designs have been put into service at the mines. Mechanical stokers have obtained only a small footing.

The more frequent use of caving systems has decreased the consumption of timber. That is one of the advantages sought. Some systems require a less number of pieces, while others save principally in the size of the timber. To the saving in the cost of the timber must be added the saving in the labor of handling it. Greater judgment has undoubtedly been used also in mines rooming with square sets in selecting the size of timber.

Without more data than is at hand it cannot be said whether much or little saving has been made on the whole in the use of explosives by the change of systems. Those plans which tend to crush the ore by excavating from below would seem to be more economical in this regard. A tendency to use explosives of higher strength may properly be noted.

When the panic came and the mines curtailed operations there was a surplus of mining material of different kinds which was drawn on until exhausted. Like the stoppage of development work, this "skinning" process could only be for a time and when expansion came again, replacements were all the greater.

The reduction in the cost of ore was aided by the reduction in the price of supplies, though not always in the same proportion. Thus, pipe fell 42 per cent, drill steel and castings 30 per cent, timber 25 per cent, coal 23 per cent, powder and cylinder oil each 20 per cent and wire rope 19 per cent. A few mines found economy in substituting wood for coal during the hard times when labor was plenty and stumpage accessible and cheap.

Such other charges as go to make up the cost did not escape the effort to reduce, but as they are peculiar to each mine, no general interest attaches to them.

In conclusion it may be said in general, that the first blow of the panic was met by a stoppage of exploring and development work and severe reductions in wages, thereafter by rigorous economy, the strictest attention to details and the overhauling of methods, and, as the outlook brightened, and wages advanced again, by labor saving plans and the increase of outputs.

Before we forget, if we can ever forget, the anxious appeal for "a job for one man" when any superfluous expenditure endangered the scant livelihood of the few at work and the occasional vacancies were filled with those whose earnings would reach the greatest number, I desire to set down, that in answer to inquiries as to their claims and the number depending upon them, I never received an untrue answer and was confirmed in the belief that our working men are, as a rule, truthful.

But even adversity has its uses, and while we were struggling with enforced economies, others were bearing like burdens and at last we awoke to the fact that the product of our labors is being sold in foreign countries and the manufactures of America are competing in the markets of the world.

THE PEWABIC CONCENTRATING WORKS.

BY L. M. HARDENBURGH.

The object of this paper is to merely outline the process used by the Pewabic company in their iron ore concentrating works, located at Iron Mountain.

The material treated consists of fragmental iron ore carried in a sandstone. The fragments of ore are rounded and water-worn, showing that they were carried for some distance and from the nature of the deposit it is evident that they were laid down along with the sandstone. The distribution of the ore through the sandstone is fairly uniform. The fragments of ore vary in size from that of a pea up to pieces weighing a couple of hundred pounds. Occasional pockets are found which contain several tons of clean ore.

The bedding of the ore in the sandstone gives the material something of the appearance of a conglomerate. There is no close contact or cementing of the ore to the containing sandstone in consequence of which when the material is crushed the ore breaks free from the rock. On this account no middle product, that is, pieces of rock with more or less ore still attached is formed.

The ore being much harder and less friable than the containing sandstone the amount of fine ore resulting from mining and the process of crushing is relatively small. In the course of mining a considerable portion of the sandstone is crushed to fine sand, as about 20 per cent of the ore is obtained by hand picking, it is necessary to remove the fine material before any attempt is made to pick out the ore.

This is done by dumping the material as it comes from the mine on a grizzly with $1\frac{1}{2}''$ openings, which allows the fine material to pass through, retaining all pieces sufficiently large to be sorted. The material passing over the grizzly is drawn on the picking table, where the clean ore is picked out and carried by means of a shoot to a pocket near the railroad track at the lower end of the mill. At the same time any barren rock is also removed.

The picking table is made from rectangular sheets of metal 12 inches wide and 48 inches long with the ends turned up. These pieces are carried on a sprocket chain. The whole forms a shallow traveling pan 48 inches wide and 13 feet long. The material is carried by this to a 14x20 inches Blake crusher. From the crusher the material passes over a ¾-inch grizzly and into a pair of 14x24 inches butted rolls. In order to keep the per cent of slime ore formed as low as possible the material is left as coarse as it is possible to handle it on the jigs. This limit has been found to be about ¾-inch opening between the rolls.

From the rolls the material goes to a three compartment revolving screen, 10 feet long, which gives four sizes of material as the over size is not carried back and recrushed, but taken direct to the jigs. The openings in the perforated plate forming the screen are 3-16 inch, 7-8 inch and 1¼ inches in diameter.

The material passing through the 1½-inch grizzly is carried to a one compartment screen with 1¼-inch openings. The material passing over this screen goes through the rolls, while that passing through is joined by that coming from the ¾-inch grizzly under the crusher and the whole is carried to a two compartment screen with 3-16-inch and ⅞-inch holes.

The material which goes through the 3-16-inch screen is carried to the jigs by the water used in screening and is distributed by hydraulic separators. Whatever passes the separators is not further treated. It consists almost entirely of fine sand and contains very little ore. The other sizes of material are also conveyed to the jigs by water and distributed by deflecting plates.

There are 18 jigs used in dressing the ore. The jigs used are made of wood and have the eccentric motion for the plunger movement. They are the ordinary Hartz jig.

Nine jigs have three compartments with screens 24x26 inches, the other nine have two compartments with 20x32 inch screens. The three compartment jigs are used for the three larger sizes of material, the two compartment being used for the material through the 3-16 inch screens.

The ore and rock separates very quickly and completely. The ore is removed from the jigs by automatic side discharges. These discharges are made by cutting an opening 8 inches wide in the side of the jig down to the screen frame. This opening is then covered with a flat "U" shaped piece which extends down to within ½ inch or so of the screen, depending on the size of ore which is to pass underneath. This piece is wide enough that it extends above the material on the screen. This opening is also provided with an adjustable plate which serves to regulate the depth of bed. Thus in discharging the ore passes under the "U" shaped piece and is crowded up over the adjustable plate.

The ore as it discharges from the jig falls into a pocket of about 300 pounds capacity fastened on the side of the jig and extending above water level. When the pocket is full the ore discharge opening is closed and a wedge shaped door in the bottom of the pocket opened, allowing the ore to run into a launder. Sufficient water follows the ore to carry it to the car at the bottom of the mill.

The hutch work from the jigs dressing the three larger sizes of material is mostly clean ore mixed with a little rock and considerable fine sand. This is conveyed by launders to the finishing jigs, two in number. The hutch work from the smaller sizes is mostly sand, but carries a little ore. This material being very closely sized the ore is removed by passing over hydraulic separators.

Four sizes of ore are produced. the bulk being between $\frac{1}{8}$ inch and $\frac{1}{2}$ inch in size. The smallest size obtained from the jigs is that over a 20-mesh screen. Much of the ore from these jigs would pass through the screens were it not for the bed of ore on the jig. The hutch, as has been stated, contains mostly sand. No attempt is made to treat the slime ore.

The only serious difficulty encountered has been the conveying of the material from the screens to the jigs. Owing to the nature of the material to be conveyed the wear on the conveying machinery was excessive. The fine sand would rapidly cut out all bearing parts. Any attempt at lubrication only intensified the difficulty. This was overcome by conveying all the material by water. With a slope of 2 inches to the foot a small stream of water conveys even the largest sizes of material.

The amount of water available for use is about 800 gallons per minute. A considerable portion of the water used on the jigs at the upper end of the mill is used again for the same purpose at a lower level in the mill. The capacity of the mill is from 280 to 300 tons of crude material per day, depending on the quality. Three men and eight boys per shift are required for the operation of the mill. The power is furnished by a 12x24 inch, 65 horsepower Corliss engine.

In giving this description of the process I have not thought it best to go into cost of dressing or per cent of ore recovered from crude material for the reason that the crude ore has been obtained almost entirely from opening work necessary for the future working of the deposit and the material thus obtained has not been sufficiently uniform to make an estimate of the saving capacity of the mill. The drifts and tunnels in the mining work were all driven in the leaner ground and it is only this leaner ore which has been treated up to the time the mill was closed down on account of inability to secure even a moderately full supply of mine labor. When the conditions are such as to again warrant the operation of the mill I shall be pleased to give the Institute more detailed information as to average per cent of ore reclaimed from material treated and such information as to cost of dressing, etc., as may be of interest.

ELECTRIC SIGNALS AT WEST VULCAN.

BY A. W. THOMPSON.

No. 2 Shaft is a strong down cast shaft and as it is wet in some places, a great amount of labor was required to cut the ice in winter and keep the pull lines in condition, so that the knockers could be rung. To avoid the difficulty, in November, 1897, two wires were run down the shaft, one for a main wire for an electric signal bell for the ore skip, and one for a grade signal bell, the two wires being grounded in the mine and at the surface.

It was considered best to put in a small single stroke bell at each level, so that when a button was pushed, the bell in the level would ring to indicate that the line was in working order. An 8-inch vibrating bell was used in the engine house to make a loud signal, but it was soon discovered that a vibrating and single stroke bell would not work successfully together as the current through them would not break in unison. The vibrating bell was taken out and replaced by an 8-inch single stroke bell, with an indicator to record the number of strokes rung.

After using the electric bells for three months in No. 2 Shaft and finding that they greatly increased the speed of hoisting, four strands of Clark's heavy, insulated wire were run in C Shaft, one for each of the three cages, and the fourth for the return wire, with 8-inch single stroke bells in the shaft house and engine house. Generally the wires leading from the shaft house to the engine house were not connected with the wires in the shaft, but could be by turning a switch at the shaft house. As this shaft is wet and there was a line of 4-inch pipe not in use, the four strands of wire were tied together every five feet and passed down the pipe.

After the bells were set working at C Shaft, two more strands of wire were run in No. 2 Shaft, one for the timber skip and the other to serve as a return wire for the three lines as there was difficulty in getting a good ground in the mine. Thereafter the elctric bells in No. 2 Shaft gave excellent results from March until November when No. 2 Shaft went out of commission.

At C Shaft the bells worked very well from the shaft house to the engine house, but from the mine to the surface it was impossible to keep the bells working for more than a few days at a time. The difficulty was that water would reach the connections or buttons, setting up either electrolysis which would corrode the wire off, or a short circuit in the buttons. In the latter case, the hammer of the bell remains down until the button is cut out or repaired.

When hoisting in No. 2 Shaft stopped the output from C Shaft greatly increased and to improve the facilities for handling the ore through it, one of the cages was taken out and the compartment divided for two ore skips. To secure the greatest speed in handling the skips, it was considered best to use electric bells and ring direct from the mine to the engine house.

Having had poor success with the bells in C Shaft up to this time, all the wires, bells and buttons were taken out and the shaft and stations re-wired, using the same wire and buttons, but leaving out the small bells in the stations to avoid making any more connections than were necessary.

The following is the system which is in use at present, as the result of an experience of over two years:

Six strands of wire are used in the shaft, as shown in the diagram. Five act as main wires and one as a return wire. The six wires are tied together every five feet, forming a cable. This is passed down through an iron pipe. The pipe is made tight by a tapered wooden plug, which is split and grooved to allow spaces for six wires. The plug is driven into the pipe and rosin melted and run into the groove around the wires, sealing the wires in the pipe. To make sure that the wires will not draw though, a clamp is put on them above the plug. At each level the wires are brought out through a tee in the pipe to connect with the buttons; then they are passed back through the tee again and dropped to the next level below. After passing the wires back into the pipe, a plug similar to the one previously mentioned, is inserted in the tee and the wires sealed and clamped.

All connections are soldered, using the best blow-pipe solder and powdered rosin. F. T. Allsop writes in "Practical Electric Bell Fittings," that "spirits should never be used on small wires as corrosion will take place and rapidly eat the wires asunder, especially in damp places." After the connection is soldered it is insulated with okenite tape and a heavy coat of Stockholm tar applied. Then a tight fitting piece of rubber tubing about four inches long is slipped over the joint and bound at each end with a small copper wire. The connections made in this way have stood for the last four months and are in first class condition. In the shaft house and the stations, for extra protection, a box-casing, painted inside and out, large enough for twelve wires is used. As soon as the wires are put in, the cover, which fits snugly, is painted and driven into place, making the joints water-tight. This puts the wires out of harm's way and makes a neat appearance.

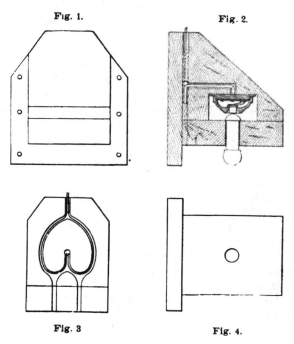

Fig. 1. Fig. 2.

Fig. 3 Fig. 4.

We have found that if the buttons are not protected, the rubber gaskets are soon worn out, the buttons become clogged with dirt and if not overhauled, a continuous current is started, which will eventually destroy the battery. As a protection to the buttons, we have devised a casing-block which prevents the water from reaching the buttons. Fig. 1 is a front view. Fig. 2 shows a cross section of the block with a water tight push button in place. The dotted lines represent the wires running from the push button to the back of the block. Fig. 3 shows the wires as they come out of the block and run around in the groove to the top of the block. The grooves are cut to the bottom of the block to allow the water that may follow the wires to drip off. Rosin and tar are run around the wires where they enter the center of the block to keep the water from reaching the button by way of the wires. Fig. 4 is a view of the bottom of the block without the spill in place. It will be seen by Figs. 1 and 4 that the back piece of the protection is made longer and wider than the block, for the purpose of fastening the block in place.

The diagram shows the system of wiring, which includes seven main wires on which are 47 buttons, and one return wire. The main wires each have an 8-inch single stroke bell with indicator attached, and are operated by four batteries. Two of the main wires run only from the shaft house to the cable enginehouse. The heavy black line represents the return wire, which is connected to the negative pole of each battery and all the bells and buttons. The main wire for the grade bell, has a button on each level. By following the main wire from the positive pole of the battery in the cable-engine house to the grade bell in the shaft house; thence to one of the buttons in the mine; thence by the return wire to the negative pole, it will be seen that the grade bell will ring if any button on this line is pushed. On the men's cage wire, there are two buttons at each station, one of which can be rung from the cage. One battery is sufficient for the two skip lines as the skips are run in balance and only one bell is rung at a time. One battery answers for the grade bell and both cable engine bells as they are never rung together.

When the system was first put in, there were only two batteries, one in each engine house on the return wire. The result of putting the battery on the return wire is that if a button is pushed on two or more lines at a time, the electromotive force on each will be much less than when one line is in use and the bells will not ring properly. If the batteries are distributed the chance of all the lines giving out at once, is practically eliminated.

The indicator was designed by ·Mr. E. Roberts, master mechanic of the Penn Iron Mining Co. The hand of the indicator is revolved by a ratchet connected by a rod to the armature of the bell. The case and hand are the same as used for steam gauges and the face is a clock dial. The hand stops when it reaches eleven and may be brought back to zero by pulling a cord. It is adjusted to register if the armature makes a quarter of its stroke. The object of the indicator is to enable the men to see, as well as hear, the signal.

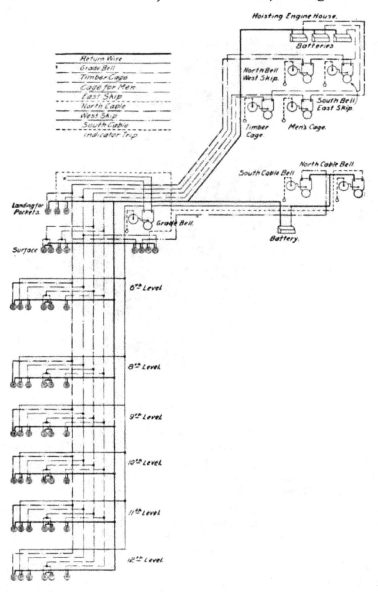

In conclusion, I would say that the secret of success with electric bells, is perfect insulation.

ELECTRIC BELL SIGNAL AND INDICATOR.

DISCUSSION ON "ELECTRIC BELL SYSTEM."

A Member—I see Mr. Roberts is here, and I would like to hear a little more about that indicator, and whether anything similar to it can be purchased.

Mr. Kelly—These indicators are not on the market, but perhaps Mr. Roberts will explain their construction.

Mr. Roberts—The pull bells used at West Vulcan before electric bells were adopted sometimes missed a stroke and the engineer could not always trust his ear. To show the strokes I made dial indicators in a rough style, but they answered the purpose satisfactorily. The bell was rung by a weight attached to the end of the bell wire from the shaft house. When the weight was pulled up it raised the lever of a ratchet which turned a wheel one notch. The ratchet

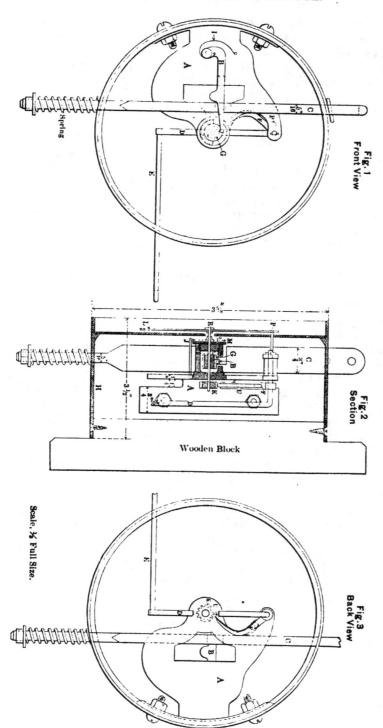

Fig. 1
Front View

Fig. 2
Section

Fig. 3
Back View

Wooden Block

Scale, ½ Full Size.

wheel was behind the dial and the spindle of the wheel carried also the pointer in front of the dial. When the weight
fell the bell rang.

When the electric bells were put in, the engineers missed
the indicators and time was lost and dissatisfaction caused
by delays in waiting for a repeating of the signal. In making indicators for electric bells it was essential to make them
delicate enough not to require much power from the magnet.
In looking over some old steam gauges I noticed on the
spindle a twelve-tooth pinion, a hair spring and a hand which
seemed to be light enough and the case also suited. Then
I needed a larger pinion with twelve teeth and this I found
in a discarded blasting battery. The hole of the pinion had
to be bushed to fit the spindle and one tooth was cut out so
as to limit its action to one revolution. After the bell is
rung eleven times the hand will not move further, owing
to the absence of the twelfth tooth.

These indicators have worked perfectly and have required
no repairs since they were made three years ago. Eight
are in use at present.

Referring to the drawings Fig. 1 represents a front view
with the hand and dial removed; Fig. 2 a vertical cross section
through the center; and Fig. 3 a back view with back removed.
H is the case; A the metal frame which carries the moving
parts. The outer end of lever E is attached to the armature
of the electric bell. A motion of 5-16 of an inch is sufficient
to operate the register. The inner end of E is jointed
to the lever D which is fixed to the curved bar F. On
the face of F and extending a little below it is a spring
U which is a particular part of the instrument. The point
of U engages in succession the teeth of pinion K. The
spindle R carries the back pinion K, the front pinion G
and the hand L. The bar B, held by the spring I, prevents the spindle with its pinions and hand revolving back-
ward, but when the trip bar C, from the top of which a cord
runs over pulleys to within reach of the engineer, is raised
it lifts B and allows the hair spring M to turn back the
spindle pinions and hand until the hand returns to the figure
0 or 12 on the dial where the pin P arrests its further motion.
Then the indicator is ready for a new signal.

A Member—I would like Mr. Thompson to explain how the
batteries are tried, as to whether the cells are all right.

Mr. Thompson—The batteries are overhauled once in three

months. Each cell is tested separately with Warner's elec-tric gauge. When connected to a new Le Clanche cell this gauge should register from thirty-six to thirty-nine units, showing the electromotive power of the cell to be about one and four-tenths volts. Cells that show twenty or more units are placed in use again. All cells below twenty and above ten units are marked and laid away to dry. All below ten units are destroyed. We do not advocate using any cell that will not show twenty or more units, except in an emergency. By washing the jars clean, dipping them in paraffin, removing the crystals from the bag elements and zinc, and renewing the salamoniac, we find that the Le Clanche cells are good for eighteen months.

Eight cells are sufficient to ring the bells when the battery is first started, but twelve are necessary after the battery has been in use a few months. As it requires from one to eleven strokes of the bell to make a signal and the signals are sent in fast, it is necessary to keep the batteries as near the maximum power as possible.

Mr. Pope—Do I understand that you give these signals from a descending or ascending cage?

Mr. Thompson—No, sir. There are two buttons at each level. One can be rung from the station and the other from the cage when the cage is standing at the level.

Mr. Knight—I would like to ask how you consider the dry-ing of the battery would renew it?

Mr. Thompson—In drying the battery the oxygen in the atmosphere acts as a depolarizer and, to a certain extent, renews the carbon in the bag or porous cup.

Mr. Knight—I would suggest you try a little hot water on them. The porous cup of a Le Clanche battery, after a time, gets a coating of crystallized salamoniac. Scraping the cell and soaking in hot water removes this scale and keeps the pores open and allows chemical action to take place between the salamoniac and carbon used, thereby renewing the battery.

Mr. Pope—Something Mr. Thompson said leads me to ask him if he has made any experiments with signals from the cage or skip. If so, where did he place the button and how was it connected with signal wire?

Mr. Thompson—There are buttons in elevators, but none that I know of in mining systems.

Mr. Denton—I would like to ask Mr. Thompson if he has any method for locating short circuits or breaks?

Mr. Thompson—If the bell should stop ringing and the hammer of the bell remain up, we would expect to find the circuit broken, though not necessarily a broken wire. Should the hammer remain down, we would look for a short circuit. In the former case, if a wire in the pipe was broken, it would be necessary to draw out the wires.

Mr. Denton—My question was intended mainly to inquire whether there is any arrangement in the pipe carrying the wires, for getting at the wires quickly?

Mr. Thompson—No, sir. As the wires run from one level to another in an iron pipe we locate the breaks and short circuits with a small bell at the buttons where the wires enter the pipe at each level. We have an extra wire to use, if any main wire should break and can connect it each side of the break where most convenient.

MINE DAMS.

BY JAMES MACNAUGHTON.

The necessity for building mine dams, throughout this part of the iron district of Michigan has been of frequent occurrence.

Especially has this been true at the Chapin Mine, where the principal shaft is located in the limestone and where the greater part of the cross-cutting to connect this shaft with the ore body is through limestone.

During the sinking of the shaft in question, viz., the Hamilton shaft no water of any importance was encountered until a depth of 1460 feet was attained, when a large water course was tapped, which filled the shaft to within 80 feet of surface in less than 36 hours.

The Ludington Mine, adjoining the Chapin, encountered water in one of the lower levels near the limestone, which entirely flooded that mine. After these two mines had been flooded for two or three years, they were bailed out, when it was found that the inflow was no greater than that previous to their flooding. This fact would indicate that the flow causing the flooding was from a large vug or reservoir, and that when it was exhausted the inflow became normal.

In the first cross-cut from the Hamilton shaft to the Chapin ore body, which is 750 feet from surface, a water course was encountered at a point south of the shaft in the limestone. This water, when first struck, had a pressure of 276 pounds to the square inch; but by drilling several holes from the breast of the cross-cut into the water course and allowing the water to flow, the pressure, after several hours, was exhausted. The breast was blasted, and it was found that the water was flowing out through a large hole or water course in the bottom of the cross-cut. The normal flow from this point after the pressure was exhausted was 750 gallons per minute. The water course was enlarged by removing all the loose rock for a depth of 18 feet, the water being syphoned out of the vug back through the cross-cut and down the shaft to the pump, which was located at a point 580 feet lower down in the shaft. A horizontal section of the hole after being thus enlarged, showed it to be about 7 feet 6 inches long by 5 feet wide. It was inclined slightly to the west.

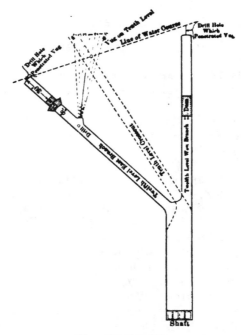

Plan of cross-cuts.

An extra heavy 10-inch pipe with 10-inch gate valve on upper end was used in the vug, through which to syphon the water. A platform was made with timber around the 10-inch pipe 17 feet below the bottom of the level. The cavity thus formed, from the platform up to the bottom of the level, was filled with a concrete composed of one part German Portland Cement to two parts of sharp clean sand and four parts of broken limestone. Great care was taken in getting good contact between the concrete and the sides of the vug. To prevent leakage between the outside surface of the pipe and the concrete, the latter was not rammed tightly around the pipe, and the small crevices thus left were filled, after each foot in height of concrete had been completed, with a mixture of equal parts of sand and cement. After the entire cavity had been filled, the concrete was allowed to set for a week before the 10-inch gate valve above referred to was closed.

The entire flow of 750 gallons per minute was thus held back and in less than 12 hours the pressure on the valve had reached 276 pounds, or the same pressure as when the water was first encountered.

On the second cross-cut from the Hamilton Shaft, which is 960 feet from surface, water was encountered at the same distance south of the shaft, but at a point a little further east than on the level above. This water, like that on the first level, was struck while drilling, and showed a pressure

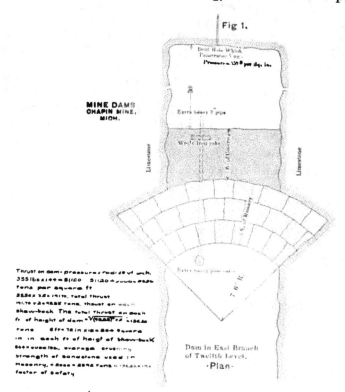

MINE DAMS
CHAPIN MINE,
MICH.

Dam in East Branch
of Twelfth Level.
·Plan·

Fig 1.

of 355 pounds to the square inch. The flow of water at this point was stopped by plugging the drill hole with a pine plug, through which was a pipe with gate valve attached. When the plug was firmly in the hole and secured by means of braces and yokes, the valve was closed.

It soon became apparent that the breast of the cross-cut was becoming weak (there being less than 2 feet of solid rock between the breast and water course); and it was decided to build a dam in the cross-cut 30 feet back from the breast to prevent any sudden inrush of water to the mine, should the breast of the cross-cut fail.

This dam was made in the shape of a circular arch on its side, the arch being 6 feet thick and having a radius of 7 feet 6 inches. On the crowning side of the arch, concrete to a thickness of 5 feet, was laid, this latter to act as a sealing device, while the stone arch provided the necessary strength.

Local sandstone was used in the construction of the arch. A strong 3-inch pipe with gate valve on outer end extended through this dam, to carry away leakage from the end of cross-cut while the dam was being constructed. The mortar used in the construction of the arch consisted of one part of German Portland cement to two parts of sand.

When the concrete and mortar had been given a reasonable time to set, the gate valve was closed, thereby accumulating the leakage in the space between the dam and breast. As soon as this space became filled with water, the dam assumed the total load due to the height of water in the vug. The total load on this dam amounts to 1840 tons, or 25.26 tons per square foot of surface exposed.

In order to avoid the necessity of penetrating the water course encountered on this level, and also to reach by a more direct route a large ore body that was not known to exist at the time of starting the original cross-cut, a new cross-cut was started by branching off from the original one at a point some distance back from the dam last above described. When this opening had reached a point about the same distance from the shaft as the original one, only 150 feet further west, water was again encountered while drilling. This supply was found to have the same pressure as that in the other cross-cut in this level, viz., 355 pounds per square inch. The efforts to stop the flow at this point did not meet with success. The breast of the cross-cut was badly broken up, many small openings or vugs being exposed, although in none of them was there any water flowing. A plug 6 feet long, with pipe and valve, was forced into the drill hole. The valve was closed and the flow stopped, but only for a short time. The plug had not reached the bottom of the hole by two feet and the rock separating the inside end of the drill hole from the exposed vugs broke away.

As it was impossible to stop the flow of water from the vugs on account of their irregular shapes, it was decided to build a dam in the cross-cut at a point 58 feet back from the breast. While preparations for starting this dam were being made, the large bailers in the Hamilton shaft were gotten ready for use. The flow of water at this point had subsided slightly, being now only 900 gallons, which in addition to the 600 gallon normal flow at this shaft made a total of 1500 gallons per minute to be pumped.

After the bailers were in readiness, an effort was made to still reduce the height of water in the vug by increasing

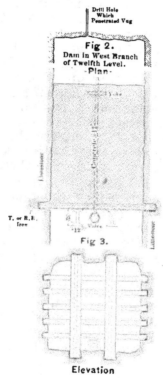

Drill Hole Which Penetrated Vug
Fig 2.
Dam in West Branch of Twelfth Level.
-Plan-
T. or R.R. Iron
Fig 3.
Elevation

the area of flow, and two other holes were drilled, both of them penetrating the vug. In this manner the flow was increased to 1600 gallons per minute from this point, or a total for the pumps and bailers of 2200 gallons. The pressure was reduced to 68 pounds per square inch. At this point both the pressure and flow suddenly increased, the former rising to 102 pounds per square inch and the latter to almost 1800 gallons per minute. This would probably indicate the existence of a second or parallel vug, which, by reason of the pressure being reduced in the first vug, thereby weakening the intervening wall, broke through into the first vug. During 12 hours the pressure gradually reduced to 80 pounds, when there was a repetition of the occurrence, the pressure rising to 127 pounds per square inch and the flow increasing somewhat, although the amount could not be determined, as the last two holes drilled were immediately plugged, and work resumed on the dam.

A temporary dam built of timber and sod was constructed near the breast. This raised the water 5 feet high and permitted its being carried in launders past the point at which the permanent dam was to be built.

All loose rock was picked from the bottom, sides and back for a length of 22 feet at the site of the permanent dam. After concrete to a thickness of 18 inches and for a length of 20 feet had been laid, an 8-inch pipe with gate valve, each capable of withstanding a pressure of 800 pounds per square inch was placed centrally between the two sides, the pipe extending about a foot outside the dam at each end. Covering the inside end of the pipe was a screen, to prevent any of the sod or material incident to the construction of the temporary dam from clogging the pipe. A heavy wrought iron yoke was used to anchor the pipe into the concrete. When the concrete had attained a height of 5 feet and had sufficiently set, the temporary dam and launders were removed, and the flow of water began through the 8-inch pipe.

When the concrete had reached a point about 9 inches from the back, it was discontinued. The remaining opening was filled with hard brick and cement mortar: first, because the concrete could not be properly rammed in so small a space; and secondly, to reduce the openings for cement as much as possible and thereby prevent excessive shrinkage.

There seemed no question but that this dam would stop the flow of water, provided it did not move bodily, the total load when applied being in excess of 2500 tons. As an extra precaution lengths of 70-pound rails were placed horizontally in the concrete with the flanges facing out and flush with the face of the concrete. The ends of these rails were cemented in hitches 6 inches deep in each side wall. Two steel grinders 13 feet long and 32 inches deep, with 12-inch flanges, were placed vertically so as to give a perfect bearing to the face of the concrete and flanges of rails. These girders were so placed as to divide the face of the dam into three spaces of 3 feet 4 inches each. The girders were cemented into hitches 18 inches deep in the back and bottom.

In less than four weeks from the time the inside face of this dam had been completed, the valve was closed, provision having been made for the air existing between the dam and breast to escape through a small pipe leading to the highest point in the back near the dam.

After all the air had escaped the dam seemed to assume the load suddenly, the hand on the gauge moving from 0 to 220 instantly.   In six hours the pressure had reached 340 pounds, and in the course of two or three days this water resumed its normal pressure.

It is a peculiar fact that the point at which water was encountered on the first cross-cut and the two points on the second cross-cut lie in the same vertical plane.

At each of the points the water when first struck carried large quantities of fine sharp sand, and although attempts were made by building temporary dams to provide for settling before pumping, a large quantity of sand passed through the pump, to which the valves and plungers can bear evidence.

ECONOMY IN THE MANUFACTURE OF MINING MACHINERY.

BY CHARLES H. FITCH.

The elements of cost in mining machinery are complex, and as sundry of them are commonly estimated as proportional to others, any change in the latter is multiplied in the aggregate like an extension of lazy tongs.

The basal elements are weight and value of materials (to which a small percentage of cost is added for warehousing), and time and rate of labor. These are obtained for every considerable portion of every job by the use of cards sent out to the shops from the office and returned with the required data. These cards are a great convenience. A pack of them girt by a rubber band is the shop record of basal elements of cost on a given job.

In the factory office clerks transcribe the card values of labor and materials upon broad cost sheets with tabular rulings. To these sheets pinned together as folios the selling correspondent resorts in fixing the figure which he asks the mining man to pay for the machinery.

These figures, however, necessarily include other costs and expenses of a general character not belonging to any particular job. Such expenses are called burdens, and are taken off from time to time to establish the ratios which they bear to the aggregate of assignable costs.

The foundry, smithy, machine shop, etc., each has its separate burden. Above these is the burden of office expense, management, advertising, selling, shipping and reasonable profits of business if any. These constitute additions to the price none of which the customer ought to escape.

In manufacture as in mining labor is the most important basal element, but its cost does not form so large a part of the value of product in manufacture as in mining.

In the foundry costs of materials, supplies and other expenses usually exceed cost of labor, and this fact makes the foundry a specially promising mine for economy of administration. Small details of buying and use assume handsome aggregates in the course of a year and should enable a pains-taking economist with proper recognition and authority to earn a good salary in many of our large general foundries, and at the same time provide a good extra profit for his employers, that is, if their competitors do not get equally clever and devoted men, economize in their foundries and then cut prices to correspond in the selling.

The mining product is staple and its price is fixed by market conditions, but the machine factory product has an elastic value, and is multiform in specification. The mine manager has a consistent and one-sided problem—to reduce cost of production, and so leave a larger margin of profit or *found* value. The works manager has a double problem, to be economical in production and also liberal to his own interest in selling. His profit is not a natural find. It is grudged him on both sides, by the workmen with whom he economizes, and by the customer who would like to economize with him, and is able to do so when competition eliminates the capital value of his betterments, and tends to leave him with no rewards but those due to a higher form of skilled labor.

Not a few highly organized shops making engines, machine tools and other fine machinery are poor dividend payers. Especially do those shops fail to pay dividends in which the stockholding interests are partly alienated. If two party interests become alienated the majority in control does not relish paying tribute on a large block of minority stock, but would prefer to divert the dividends to betterments, and large salaries to preferred persons in executive places. With such persons dominating the office end of a works, liberality rather than economy is the rule of service. Privilege and favor leak down; sinecures are allowed and become patterns to subordinates. The enthusiasm for economy is impaired by practices distinctly opposed to it, and held in higher favor. It is seen that there is not one rule for all, and the effort for economy becomes more perfunctory than strenuous in those of whom it is expected.

The manufacturer locates himself in a nest of accumulating expenses which build about and hinder his growth. If he expands to employ a city he is taxed on city values, and must pay unearned increments out of earnings. He has not the bulwark and advantage of the mine operator in possessing a bank of found value to draw upon. To share this advantage the factory must enter into partnership with the mine, and in such a union it is the mine which will have the capitalistic and potential preponderance. Although one man or company may control both, it is the mine which will in effect employ the factory, and pay dividends. This will happen when factory processes are completely economized, and there is no liberal margin left in the business.

Systematic economy is not easy to realize. In the large manufacturing shop, building to contract, character of work is varied and changing. With continual change of conditions there is difficulty in making comparative checks upon efficiency. On the other side the selling is influenced by favor and personal prejudice, by wide variations in specifications and in claims of merit, specious or real, and by premiums on time delivery due to the improvidence of users of mining machinery. Among these vicissitudes system and economy are evasive, are not easily corralled.

Years ago I was sent by an eminent professor with a schedule of information to be obtained from an eminent manufacturer. The comment of the latter was: "Scientific men are fools. They think that there is system in these things when there is no system." I could not quite accept this dictum, but I think it will be allowed to have some force by gentlemen who do their shop work in shops, and not in magazine essays.

Despite its difficulties of attainment the manufacturer pursues, and is obliged to pursue systematic economy, and making virtue of necessity proclaims the customer to be a sharer in his saving. But if as would appear, the competitive chasing of such economy tends to thin his profits to the breaking point it is a pursuit akin to the "Hunting of the Snark," in which Lewis Carroll satirizes all forms of futile and impracticable propaganda. One of the hunters of this mysterious creature is described as finding it, and simultaneously disappears, whereat the poet observes:

"But beware if the Snark be a Boojum, for then
"You will swiftly and silently vanish away,
"And never be heard from again."

With this peril in view, if the contagion of *economy* is to infect our profits we do not care to pursue it with inconsiderate zeal.

But the mining man is expected to have no such scruples in behalf of the manufacturer. Let him look to his own economic interests, now much neglected. His societies can exist to no better purpose than to specify that which is finest for his purposes, avoiding improvidence, the heresy of cheapness which is not economical, and ancient and freakish features of machinery which are not to the best purpose.

If the first cost of what is best is now formidable it is largely due to the improvidence of the mining man. Let him essay the task of specifying the best, and making his specifications standard. Then the manufacturing effort now spent in catering to meaningless and uneconomical variations will be confined to more uniform lines. Manufacturers can and will then produce the best at much less cost than now for two reasons:

1. There will be removed an element of uncertainty in the selling, which now greatly increases prices.

2. The problem of shop work will be presented in its most advantageous economic form, namely: The repeated production of one kind of work to which a man can give his undistracted attention until it is unerringly accomplished in the best way.

DISCUSSION ON ECONOMY IN THE MANUFACTURE OF MACHINERY.

Mr. Kelly—As the depth and size of mining operations increase, their success depends more and more on the solution of the mechanical problems that arise and the subject of the paper that has just been presented is of very great importance. The majority of the members of the Institute are best acquainted with the uses of mining machinery so that this paper is of special interest to them because of its point of view.

In many cases the purchaser looks to the manufacturer as an adviser with special and broader experience and the latter is in some measure responsible for the selection of the design and the rejection of what is unsuitable and antiquated. That this responsibility is appreciated is illustrated by an incident in the speaker's experience. A heavy pressure pump of large capacity was required and the agent of the manufacturer quoted a price on a triple expansion pump to meet the conditions of capacity, lift, steam pressure, etc. Triple expansion mine pumps had not been long tried, the steam pressure was not very high and a price on a compound pump was asked for. But the agent replied, "I will not make you a price on a compound pump; I cannot recommend it for the conditions. The economy of a triple pump will not permit you to buy a compound pump." The advice was accepted and the service of the triple pump shows its wisdom. It is evidently not for the interest of either party and it can hardly be said to be customary, for the purchaser to prescribe detailed specifications without consultation with the manufacturer.

Mr. Hood—I notice that the author names at least two methods of reducing the cost of machinery, one on his side of the fence and one on ours. He speaks first of economies in the foundry, of which I know something. There is no part of machinery where the product is liable to such variations as in the foundry. The difficulty of retaining the best men in this line of work is the reason for much inferior foundry practice.

The other question, that of having the members of this Institute select and specify that which is best, I am of the opinion that there would be a great deal of difficulty in getting together on a good many of the most important lines. Asking one member as to his opinion of the construction of a certain connecting rod, he condemned it, and the next gentleman asked could not praise it too highly. Each had long experience. Where the conditions vary so widely it seems to me difficult to get together on very many lines.

Mr. Channing—I fear that too much uniformity in the design of mining machinery would tend to retard improvement. There is more individuality in the design of an air compressor than in that of an angle bar. It is easy to make a trust out of the manufacturers of structural material while it would be impossible to bring under one head the makers of mining machinery.

The air compressor of ten years ago is today an inefficient machine and the improvements have come not so much from the manufacturer's desire to change his patterns as from the demand of the mining engineer that a higher speed and consequently smaller and cheaper machine shall be furnished him which shall at the same time show greater efficiency.

The usual data furnished the manufacturer simply demands so many cubic feet of free air per minute under certain conditions of altitude and steam pressure. It is for him to submit his design, price and guarantee of economy. If he sticks to a standard design he may find a competitor taking the order.

While we may have standard sizes for compressors, as the air presure, at least in mining work, varies between small limits and the steam cylinders need but vary according to the boiler pressure; yet, in hoisting engines there are so many variables that as long as ore deposits vary in their dimensions and character, fuel in its bulk and price, water in its solid constituents and abundance and labor in the efficiency, so long must the design of a hoisting engine vary.

No doubt certain sizes and types of drum might be adopted and these in combination with various standard cylinders might aid the economy of manfacture.

I lately asked for bids on three blowing engines to be used in furnishing blast for three copper smelting blast furnaces.

The pressure, two and a half pounds to the square inch, was lower than is usual in blowing engines and I was obliged to decline the proposals of those firms who offered me an engine whose blowing tubes were mere duplicates of those acceptable for higher pressures. For small hoisters and for most kinds of pumps I think we have adopted standards but for larger machines there is a constant demand for individuality and the designer of genius is bound to break away from standards and preconceived notions.

Mr. McKee—I think it would be very difficult to standardise mining machinery. With pumps the best that could be done would probably be to make steam ends of certain powers and water ends of certain capacities so designated that they may be interchanged. It is impossible to make the pumps of uniform size as it is not possible to regulate the amount of water made by a mine or the height to which it is to be lifted.

As to boilers those now built by reputable builders (and no others should be bought), conform pretty closely or should, to Hartford standard specifications. I would suggest that, as these specifications are so easily obtained, that they be secured and used where new boilers ought to be purchased.

As to hoisting machinery, it would probably be easier to standardize than almost anything else. But even this would be very difficult under present conditions. At present no two mines carry the same steam pressure or hoist the same load at the same velocity. The mine managements might get together and adopt some standard sizes of skips, but at present you will see all dimensions of skips from about one to five tons.

Here again, however, we meet the difficulty of mines of variable depth and conditions requiring hoisting at widely varying speeds. I think on the whole that it would be quite difficult to standardize the machinery at all.

Mr. Pope—I think the manufacturer of machinery does not usually consider the fact that a mining manager is very seldom able to equip his plant in a way he would like to equip it, or that he cannot tell at any time what he is going to want in the future. He is answerable always to the home office, the executive of which is usually unacquainted with mining problems and sometimes thinks he is doing great work for the company by opposing the suggestions of the manager, who is compelled to go to as little expense as possible, and if too enterprising may be removed from his position. The manager cannot disclose to the manufacturer his reasons for not buying a plant

7

which would be needed for an assured future, and therefore
the manufacturer finds fault with him, says he does not know
what he wants, and blames him for not doing what he is not
allowed to do.   I do not think the manager should be obliged
to furnish all the details of the machinery.   He can give the
steam pressure, speed of hoist, depth of shaft, and many other
essentials which he has learned the need of in his experience
and then ask for details and specifications from the manufacturer for consideration—because no two mining propositions
are alike, and the manager must adopt such machinery as his
conditions require.

Mr. Davidson—I think Mr. Pope's ideas of the subject are
correct.   The equipment of a mine cannot be determined in
advance as the equipment of a factory can be, but in most
cases, a mine superintendent must add to his machinery as his
property is developed.   As a result, we often see mines equip-
ped with machinery that is not by any means the most desirable or economical, but before one is justified in throwing it
out, he must be quite sure that he has ore enough in sight to
warrant the cost of purchase and installation of new machinery.

Mr. Fitch—Of course engineering conditions which determine approximate sizes and capacities cannot be ignored, and
mining problems must often be attacked with limited re-
sources.   Still the work of design may be approached on
broader lines reaching an improved system of graded sizes,
both in entire machines and in their elements.   Taking the
whole common stock of mining equipment the differences of
design in machinery for like uses are largely unessential.   To
that extent they may be eliminated with manifest economy in
manufacture, repair and use.

Also with every agreement upon a unit such as suggested
by Mr. Channing in the matter of skip sizes the elements of
necessary variation are reduced to a more manageable series.

Most first designs are at fault in simplicity, and a very large
machine attacked as a problem by itself is a first design, usually an expensive monument with needless efforts of invention.
A good course is to design broadly for a series of sizes at once.

Some years ago I was being shown through the Baldwin
Locomotive Works by the late Dr. Williams and expressed sur-
prise at the great variety in styles and sizes of engines which
they built.   He said: "You would be more surprised at the
small number of patterns from which we build them."   The
Baldwin people not only build more engines than any other

works but they are of the finest types, and are built for quick delivery, and require only one-half to one-third the hours of manual labor needed for similar work in large foreign shops. The reason is simple. They put brains into their business and standardize their work in every detail. This is perfectly practicable, but too sparingly practiced in our mining machinery shops.

METHOD OF MINING AT THE BADGER MINE, COMMON-WEALTH, WIS.

BY O. C. DAVIDSON.

The ore deposit afterwards known as the Badger Mine of the Commonwealth Iron Company was first found in April, 1891. The ore was discovered by test-pitting, which also showed that the drift above the ore was from eight to twenty feet and averaged about twelve feet. We were anxious to secure as large a product that season as possible, as ore was in good demand, and it was decided to strip the ore body, and at the same time start sinking shafts.

The ore body, as far as we had traced it at that time, was a little over five hundred feet long and ninety feet wide at its widest point. Three single-skip shafts were started in the foot wall, the distance between them being two hundred and fifty feet. The east shaft, owing to the position we were compelled to locate our railroad tracks, would have proven inconvenient if sunk on an incline, and this shaft was, therefore, made vertical. The other shafts were sunk at an angle of sixty-eight degrees from the horizontal, and this we afterwards found to be very near the actual dip of the foot wall.

The 72,435 tons of ore shipped the first season came from the 80-foot level, and during the next winter the shafts were sunk one hundred feet deeper, and the mine was worked as an open pit to a depth of one hundred and eighty feet, the ore being obtained by the milling method, and by under-hand stoping. In order to work the mine as an open pit to a depth of one hundred and eighty feet, and to secure all the ore on the hanging side, it was necessary to hoist a great many thousand tons of the red slates forming the hanging wall. This is a portion of the work that we do not refer to with any pride, as we now realize that it would have been more economical to have stopped open pit work at a depth of one hundred feet, and work from that depth as we did below the 180-foot level.

The next level was opened up ninety feet below the bottom of the open pit. A drift was driven in the foot wall connecting the shafts, and from this drift were driven cross-cuts every sixty-five feet. Directly above these main cross-cuts, sub-cross-cuts were driven at a depth which, when worked out, would give us a stope seventy feet high and leave a back or floor of ore above us twenty feet thick. These main cross-cuts and sub-cross-cuts were then connected with raises, which were placed at such points as would best enable us to separate the three grades of ore that were mined. Chutes were built in the bottom of these raises and the ore was mined by the milling method precisely as it is in open pits, except that we left pillars across the ore body twenty-five feet thick, and worked the ore out in stopes forty feet wide and seventy feet high. The mine was opened up and worked in this manner, the levels all being ninety feet apart until a depth of five hundred and forty feet was reached, but at this depth, the merchantable ore was practically pinched out. There was also a decided pitch of the ore body to the west so that the East shaft was only carried to a depth of two hundred and seventy feet, and as the ore in the east end of the deposit pinched out, the rooms were filled with rock from the hanging wall, mills being built up as filling progressed. When the bottom room was nearly filled, the floor above it was mined out and sent down through the mills, and after the floors were all taken out, the work of mining the pillars was begun. We first tried undercutting the pillars, but this was not at all successful as the pillar would not settle as it should, and rock from the adjoining filled rooms would come in where it was not wanted. We, therefore, began top slicing, and have found this very satisfactory, both in the matter of cost of ore and because practically all of the ore is secured.

The product of the Badger Mine up to the first of the present month is 1,368,703 tons of ore. We think the method of underground milling which we adopted enabled us to secure a larger product per man per day than could have been secured any other way. A great deal of drifting is required with this method, but though the ore is tough and does not require any timbering, it does not drill hard, and the cost of drifting is, therefore, very moderate. Very soft ore bodies or stratified or seamy ores could not probably be worked with safety in this manner, but as an evidence of the physical character of the Badger Mine ore, I would say that we had several stopes seventy feet high two hundred feet long across the vein and forty feet wide, and in one place we mined out two floors before filling, thus leaving a room two hundred and seventy feet high. It was our custom to confine our winter work as much as possible to opening, stock-piling as little ore as possible, and then when the ore could be loaded direct into cars, increase our force. In this way, we are able to attain a large product during the summer, our maximum product being 38,000 tons for one month.

BALANCING BAILERS.

BY WILLIAM KELLY.

The simplest and most effective way to balance bailers in a shaft with two compartments is to attach them to a hoisting plant of which the rope winding part is driven directly by a pair of reversible engines There are a number of "first motion plants" that fill the requirements in the Lake Superior district but they are not always available when bailing is required. Orders for unwatering mines and emergencies in pumping usually come with little warning and appliances at hand must be utilized. The advantage of balancing the bailers is so great that it is desirable to use some device for the purpose even when a direct acting plant with reversible engines is not at hand.

Some years ago two bailers were made for the West Vulcan Mine to be used should an emergency arise, and as there is a "first motion plant" there. the weight of the bailers was a secondary consideration and they were made of one-quarter inch steel.

The bailers were not used until it became necessary to unwater the Curry Mine in the early part of 1899. The hoisting plant consists of two 6-ft. independent drums on a shaft driven by a non-reversing Corliss engine. The safe load for the plant was given by the builders as 8,000 pounds. The shaft has two compartments. The plan for balancing the bailers adopted was counter-balancing each bailer separately. To determine the size of the bailers and the quantity of water that could be raised at a lift, it was necessary to deduct from the safe load for the engine the weight of 600 feet of 1 in. rope, at 1.58 pounds, 948 pounds; weight of 600 feet of 5-8 in. rope, at .62 pounds, 372 pounds; weight of bailers unbalanced to overhaul rope to engine house, 300 pounds; total, 1,620 pounds; leaving the amount of water load, 6,380 pounds. This is equivalent to about 765 gallons and as the bailers are 39 inches in diameter they were made 12 ft. 4 in. in length above the valve. The conterbalances used were 12 in. Cornish pump plungers. Scrap was put inside and the weight adjusted so as to counterbalance the bailers except so far as was necessary to allow them to start down freely from the top. The advantage of this plan is that the bailer that is at the bottom may be started just as the upper one gets to the dump and while one is being raised the other one is emptied and lowered into the water, thus saving the time it takes for emptying and filling, and keeping the engine hoisting continuously. The disadvantage of the plan is that where provision has not been previously made a road for the counter-balance has to be put in as unwatering proceeds, a matter which takes considerable time.

Just as unwatering at Curry was nearly completed another plan for balancing bailers was suggested and seemed to offer considerable advantages, and when it was decided to unwater the East Vulcan Mine the new plan was tried. The hoisting plant at East Vulcan consists of two 10 ft. drums in tandem driven by a non-reversible Corliss engine. The bailers were balanced by means of a third rope running from one bailer up over a pair of head sheaves and down to the other bailer. The feasibility of the plan depended on whether the brake of the drum of the descending bailer could be eased off so as to throw the weight of the bailer upon the counter-balance rope and yet not to permit the hoisting rope to get too much slack. The brakesmen thought this could be done. It was found in practice, however, that the brakesmen were unable to keep the two drums so nearly at the same speed and the weight of the descending drum was thrown suddenly from the drum rope to the counterbalance rope and back again with the result that in a few hours the counterbalance rope was broken.

The capacity of the hoisting engine as figured by the makers is a little over 15,000 pounds. One and one-eighth inch ropes are on the drums for which the safe working load is 16,400 pounds. The counterbalance rope is 5-8 in., for which the limit is 5,400 pounds. The weight of 880 feet of 1 1-8 in. and 5-8 in. rope is 2,300 pounds, which deducted from the capacity of the plant, gave 12,500 pounds as a safe load of water. The bailers were lengthened so as to be 24 ft. 4 in. above the bottom of the valve and being 3 ft. 3 in. in diameter, held a little over 1,500 gallons. They weighed empty, one of them 5,250 pounds, the other 5,400 pounds.

When it was found that the counterbalancing rope from one bailer to the other was an unsatisfactory arrangement, the following plans were open: First, to take out one bailer and use the compartment for a counterbalance, so as to continue the use of a bailer of full size. The size of the hoisting compartment did not permit putting in a counterbalance for a bailer in the same compartment. The second plan was to cut down the bailer so as to bring the weight of the bailer and

load within the capacity of the engine.   This would reduce the
weight of water hoisted per load nearly one-third.   A third
plan was to run the bailers as they were without any counter-
balancing device and take the chances on the machinery, but
the load was one-third more than recommended by the makers
of the machine, and the idea did not commend itself.   A fourth

plan was devised and to study its feasibility, a model was made
by Mr. Roberts and this was used to illustrate the topic in presenting it to the Institute.   The device adopted consists of a
third rope the ends of which are wound upon the two drums in
the reverse direction from that of the hoisting ropes and to
permit of the traverse upon the drums, the rope instead of

going directly from one drum to the other is carried about 200 feet to a turn sheave, as illustrated by a photograph of the model. The mode of operation is as follows: When a bailer gets to the top, the brakesman sets the brake and at the same time the brakesman at the other drum sets the brake for the bailer that is in the water. As soon as the bailer at the top is empty, which takes 12 to 15 seconds, the brakesman releases the brake of the upper bailer and its weight is carried by the intermediate rope to the other drum tending to wind the latter up. The second brakesman then releases the brake and puts on the friction and hoists the lower bailer to the top. It has worked quite successfully.

DISCUSSION ON BAILERS.

Mr. Brown—Is there a chance of getting that weight on the auxiliary rope, or does that act itself?

Mr. Kelly—Yes, sir. The weight of the empty bailer at the top is transferred through the auxiliary rope to the other drum when the brake is released.

Mr. Brown—How about adjustments?

Mr. Kelly—It is necessary from time to time to take off a round on the drum. That takes about fifteen minutes, but we do not get down thirty feet every day.

Mr. Cole—If the hoist, to begin with, had been supplied with reversing links, and operated by a lever applied from the platform, would it not be possible for one man to operate the plant? Would he not be able to operate that plant so as to run in balance and make all adjustments at any time? Is it necessary with this scheme to have two brakemen on that platform?

Mr. Kelly—I do not know that it is necessary to have two brakemen. We put on two because it was safer, and we wanted them to hoist twelve hours, and have not attempted any economy at that point. The plan suggested by Mr. Cole is feasible. On the plant we have, in addition to the reversing gear, it would be necessary to put a brake on the fly wheel in order to stop quickly, and to make the design complete there should be a small starting engine to avoid the difficulty of centering.

A Member—With the difference in the size of the ropes is there not a differential pulley?

Mr. Kelly—No, sir. There is no differential pulley because one of the drums is always free and is not required to travel at

8–R

the same speed as the other. If the ropes were of different size and the two drums revolved at the same speed there would be a differential pulley.

Mr. Cole—I can readily see that the device is very ingenious, and for you at the time was well adapted to your requirements.

After getting the mine unwatered, have you used that arrangement for hoisting ore, or for other purposes.

Mr. Kelly—We have not got to that point unfortunately, Mr. Cole, but if we had conditions where we could use both compartments for hoisting ore we would not hesitate to adopt it, particularly if there was but one level. In case there was more than one level it would be necessary for the skips or cages to make a complete trip to the bottom every time as there is no facility for quickly changing the length of the rope.

Mr. Cole—And you think that one man on the brake platform would be able to perform the duties; that is, in ten hour shift?

Mr. Kelly—I think so.

Mr. Pope—What effect would it have to use an endless rope on horizontal drums passing over two or three times in and out, forming a figure eight, without any auxiliary,—the rope to be put on, one end at the top, and the other at the bottom, passing around both drums.

Mr. Lawton—Mr. President, I do not think that that plan would work, unless grooves in the drums were so the rope could travel over and back. The rope would tend to spread.

Mr. Pope—We applied that principle some thirty years ago on a gravity road and it worked well. We used, at that time, manila rope, but probably it would not answer with wire rope.

Mr. Kelly—A single rope wound around both drums with the ends attached to bailers, skips or cages is undoubtedly feasible with such a plant if there are no grooves on the drums or the drums are grooved reversely or if the rope instead of going directly from one drum to the other passes from one drum to a turn pulley and back to the other drum to allow for the traverse.

CPSIA information can be obtained
at www.ICGtesting.com
Printed in the USA
BVHW080923231118
533806BV00010B/170/P